The Sustainable Plate: Unveiling the Science of Meat Substitutes and Impact on Global Health

Subhasree Ray • Shoba Suri

The Sustainable Plate: Unveiling the Science of Meat Substitutes and Impact on Global Health

Springer

Subhasree Ray
TVS Motor Company
Hosur, Tamil Nadu, India

Shoba Suri
Observer Research Foundation
Delhi, New Delhi, India

ISBN 978-3-031-94603-5 ISBN 978-3-031-94604-2 (eBook)
https://doi.org/10.1007/978-3-031-94604-2

This Springer imprint is published by the registered company Springer Nature Switzerland AG
The registered company address is: Gewerbestrasse 11, 6330 Cham, Switzerland

Contents

Chapter 1
Introduction

1.1 Overview

The intersection of environmental sustainability and human health has become a top priority in modern dietary choices. *The Sustainable Plate* is a book that aims to explore the relatively new territory of meat substitutes, unravelling the intricate science behind them and their significant impact on global health. As the world's population grows and environmental challenges escalate, this book offers solutions to provide sustainable nutrition for the growing population. The authors' goal has been to demystify the often-complex landscape of meat substitutes as stewards of this scientific exploration. The urgency of addressing the science of meat substitutes and their impact on global health arises due to the increasing challenges posed by traditional meat production. From deforestation to water pollution and significant contributions to greenhouse gas emissions, the environmental toll of meat production is unsustainable. Meanwhile, the growing global demand for protein necessitates alternative, more sustainable sources. *The Sustainable Plate* addresses this need by providing a well-researched, evidence-based exploration of meat substitutes. Understanding the science behind these alternatives is crucial for fostering a sustainable and healthy future in a world where dietary choices can have far-reaching consequences.

The original version of the chapter has been revised. A correction to this chapter can be found at https://doi.org/10.1007/978-3-031-94604-2_11

S. Ray, S. Suri, *The Sustainable Plate: Unveiling the Science of Meat Substitutes and Impact on Global Health*, https://doi.org/10.1007/978-3-031-94604-2_1

The narrative begins by examining the underlying scientific foundations, delving into the composition and technological intricacies of these substitutes. A thorough understanding of the nutritional equivalences and differences between traditional meat and substitutes forms the foundation of this exploration. By deciphering the biochemical intricacies and nutritional profiles, readers can make informed dietary decisions in pursuit of individual and planetary well-being.

It focuses on the impact of diet on the environment, particularly meat production, a resource-intensive practice with ecological consequences that need to be reevaluated considering sustainability imperatives. We examine the comparative impacts of meat substitutes and conventional meat production through life cycle assessments and comprehensive environmental footprint analyses. The ecological intricacies are exposed, and the potential of meat substitutes to mitigate the environmental toll becomes a critical theme, underscoring the urgency of adopting sustainable protein sources. *The Sustainable Plate* extends its gaze beyond the individual to the global stage. In a world facing food security challenges, resource depletion, and escalating health crises, the ramifications of our dietary choices are magnified. This book explores the interconnectedness of personal nutrition with broader global health perspectives. The potential for meat substitutes to address malnutrition, enhance food security, and contribute to sustainable development becomes a linchpin in our exploration, highlighting the integral role that dietary choices play in shaping the health of our planet. We recognize the imperative for a nuanced, evidence-based understanding in weaving this narrative. The call for sustainable nutrition is not a mere echo in the chambers of dietary trends but a resonant chord that harmonizes personal health aspirations with planetary well-being. *The Sustainable Plate* is more than a compendium of facts; it is an immersive journey into the scientific foundations of meat substitutes. It equips readers with the knowledge to make choices that echo through the annals of global health and environmental sustainability. Join us on this journey towards a healthier self and a healthier planet as we unveil the science behind the sustainable plate.

1.2 Why Is There a Growing Interest in Alternative Meats?

The meat industry is expanding alongside the growing global population [1]. However, the demand is expected to increase even more due to the rising gross domestic product per capita in developing countries. The meat market is projected to grow from $897 billion in 2021 to $1354 billion by 2027 [2]. Moreover, the meat market is anticipated to reach 374.4 million tons by 2029, showing a Compound Annual Growth Rate (CAGR) of 1.1%. To meet the growing need for cheap protein, the livestock industry has become more industrialized.

In the last 50 years, the production of meat has increased fourfold [3]. During the 1960s and 1970s, Europe and North America were the primary producers of meat, but presently, Asia has taken the lead and accounts for over 100 million tons of meat produced annually out of a global total of nearly 300 million tons. The production

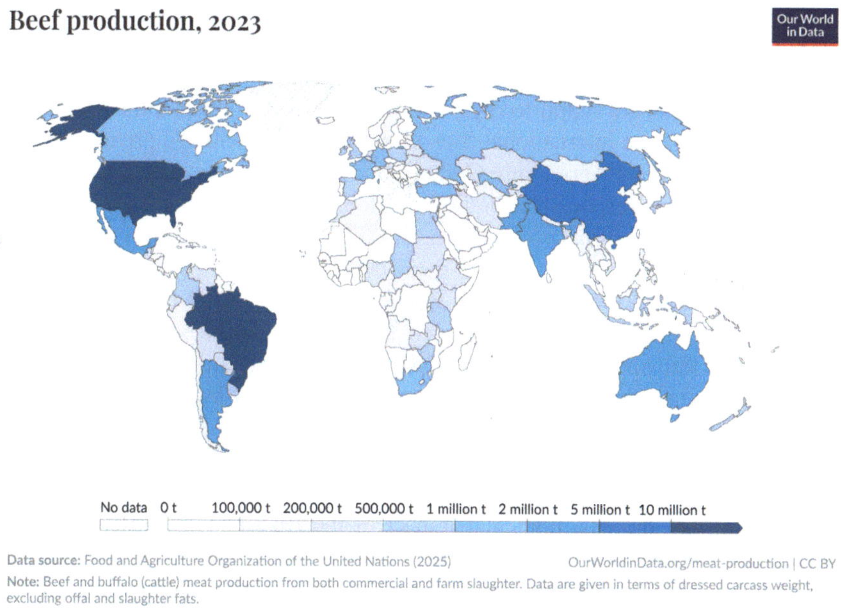

Fig. 1.1 Global beef production, 2023. (Source: Our World in Data. https://ourworldindata.org/grapher/beef-and-buffalo-meat-production-tonnes)

of poultry and pig meat dominates the current meat production. The per capita meat consumption has also significantly increased, with the average person consuming around 43 kg of meat in 2014, compared to 20 kg in 1961 globally. Figure 1.1 illustrates the global beef and buffalo (cattle) meat production in 2023 from commercial and farm slaughter [4].

Increased meat production uses almost half of the habitable land and contributes to over a quarter of global greenhouse gas emissions, while it produces less than 20% of the total calories and less than 40% of the total protein. Therefore, it is highly inefficient. The production of beef has a significant impact on climate change as it emits greenhouse gases (GHGe), such as methane, nitrous oxide, and carbon dioxide [5]. Studies indicate that ruminant livestock are responsible for approximately 7–18% of methane emissions caused by human-related activities on a global scale. Greenhouse gas emissions are measured in kilograms of carbon dioxide-equivalents. Non-CO_2 gases are weighted based on the amount of warming they cause over a 100-year timescale. Poore and Nemecek (2018) reported that beef, mutton, and lamb production top the chart in emitting potential GHGs [6]. These emissions account for the global warming potential of all greenhouse gases over a 100-year timescale, not just CO_2. Among food items, beef from beef herds generates the highest emissions at 99.48 kg CO_2eq/kg, followed by lamb and mutton (39.72 kg), and beef from dairy herds (33.3 kg). Other high-emission foods include farmed prawns (26.87 kg), cheese (23.88 kg), pig meat (12.31 kg), and poultry (9.87 kg). In contrast, plant-based foods generally produce significantly lower emissions; for instance, eggs (4.67 kg), rice (4.45 kg), milk (3.15 kg), tomatoes (2.09 kg), maize (1.7 kg), and wheat and rye (1.57 kg). The lowest emissions are observed in peas (0.98 kg), bananas (0.86 kg), potatoes (0.46 kg), and nuts (0.43 kg). These findings that are supported by the Livestock, Environment and Development (LEAD) Initiative, an international consortium of government and private agencies based at the Food and Agriculture Organization (FAO), who revealed that the livestock production is responsible for 18% of global greenhouse gas emissions. This includes 9% carbon dioxide emissions and 37% methane gas emissions worldwide [2].

Rowe et al. (2022) noted that New Zealand's sheep industry exports around 20 million lamb carcasses annually, which are mainly produced from grass [7]. However, the fermentation of forage by grazing ruminants emits methane, which is responsible for one-third of all greenhouse gas emissions in New Zealand. According to a recent study conducted by Kirwan et al. (2023), in Ireland, the median greenhouse gas emissions were 2.77, 2.93, and 4.31 kgCO_2eq, and freshwater use per day was 88, 144, and 307 L for children, teenagers, and adults, respectively [8]. The study found that the environmental impact of the Irish population exceeded the planetary boundary for GHGe by at least 148% for all population groups, meaning that the amount of greenhouse gas emissions produced by the population was significantly higher than what the planet can sustain. The study also found that meat and meat alternatives (27–44%), eggs, dairy, and dairy alternatives (15–21%), and starchy staples (10–20%) were the main contributors to GHGe. Meat, dairy, and grains were identified as the main contributors to cropland, nitrogen, and phosphorous use (79–88%). Clark and colleagues conducted a study in 2022 where they applied NutriScore to evaluate 57,000 food products in the United Kingdom and Ireland [9]. Their findings revealed that food products could be categorized according to low (such as sugary beverages, fruits, breads), intermediate (such as many desserts and pastries), or high in terms of environmental impacts (such as meat, fish, cheese). A study conducted by Sranacharoenpong et al. in the USA in 2015 found that producing 1 kg of protein from kidney beans required significantly less resources than producing 1 kg of protein from beef [10]. Specifically, it required 18 times less land, 10 times less water, 9 times less fuel, 12 times less fertilizer, and 10 times less pesticide. In comparison to producing 1 kg of protein from chicken and eggs, beef generated five to six times more waste (manure) to produce the same amount of protein. The study concluded that replacing beef with beans in meal patterns could significantly reduce the environmental impact worldwide and should also be encouraged to help reduce the prevalence of non-communicable chronic diseases.

Along with the GHGe, ill-treated disposal of cattle production waste causes water pollution. Poor grazing management leads to sediment contamination of surface water and groundwater. Beef production is water-intensive, mostly used for growing cattle feed. With the expansion of the global cattle industry, the beef slaughter and leather industries have also grown rapidly. The waste generated from these industries, rich in organic matter, heavy metals, and caustic solutions, poses threats of pollution if treated inappropriately. Additionally, livestock farming is a

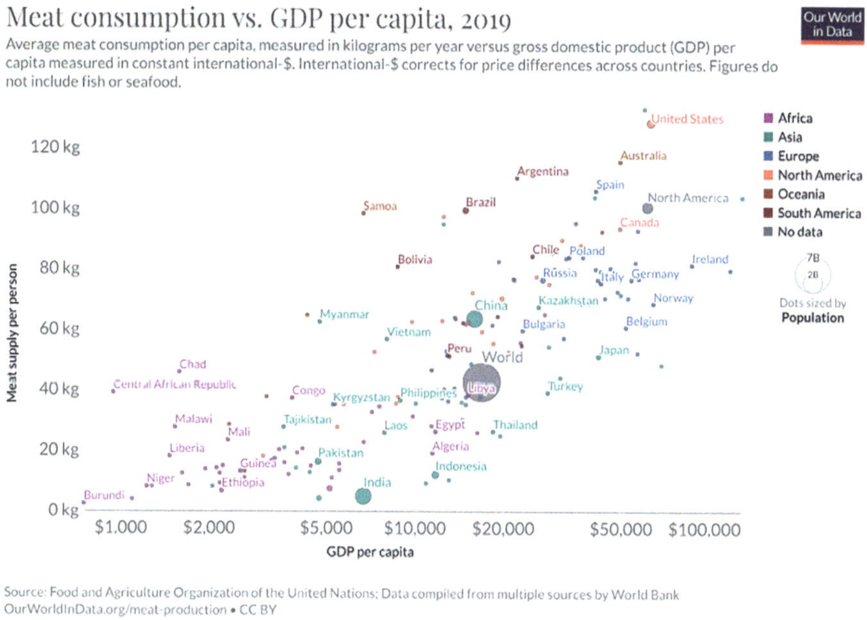

Fig. 1.2 Meat consumption vs. GDP per capita, 2020. (Source: Wikimedia Commons, This file is licensed under the Creative Commons Attribution 4.0 International license. https://commons.m.wikimedia.org/wiki/File:Meat-consumption-vs-gdp-per-capita_%28OWID_0668%29.png)

major contributor to soil erosion worldwide. Turning forests into pastures, overgrazing, or using marginal lands for feed production can result in losing topsoil and organic matter, which can take decades or even centuries to recover.

It is widely observed that a person's income has a significant influence on their meat consumption habits, especially when comparing different countries. Figure 1.2, a scatter plot derived from multi-country per person meat consumption data by FAO and GDP data by the World Bank, visualizes the relationship between the per capita meat supply (y-axis) and the average GDP per capita (x-axis) [4]. The plot reveals that in the year 2020, there is a positive correlation between a country's wealth and the amount of meat consumed by its people. In other words, people in richer countries tend to eat more meat than those in poorer countries.

The consumption of meat has been on the rise, leading to an increase in health problems such as heart disease. Yan et al. conducted a study in 2023, which involved 204 countries and territories between 1990 and 2019. The study aimed to assess the burden of Ischemic Heart Disease (IHD) caused by a diet high in red and processed meat in these countries during this timeline. Data from the Global Burden of Disease (GBD) 2019 was used for this purpose [11]. The study showed that globally, a high intake of red meat was responsible for 351,200 deaths from IHD in 2019, while a high intake of processed meat was associated with 171,700 deaths from IHD. Although the corresponding age-standardized rates declined between 1990 and 2019, the numbers of deaths and disability-adjusted life years (DALYs) increased. According to this study, in 2019, China had the highest number of deaths caused by a high intake of red meat, while the United States of America was associated with the highest number of deaths caused by a high intake of processed meat for IHD. Meat consumption is linked to harmful gut health outcomes. A randomized control trial conducted by Gao et al. (2021) found that fried meat consumption disrupted glucose regulation, elevated endotoxin levels, and triggered inflammation by altering gut microbiota and microbial-host cometabolites [12]. Landry et al. (2023) reported that after 8 weeks, the 22 participants randomized to a vegan diet experienced significant decreases in LDL cholesterol concentration, fasting insulin level, and body weight compared to those who followed an omnivorous diet [13]. However, the evidence associating negative health outcomes of red meat is contradictory. The US Department of Agriculture's Nutrition Evidence Systematic Review (NESR) team conducted a meta-analysis of 21 relevant RCTs to investigate the association between red meat intake and type 2 diabetes mellitus (T2D) [14]. The authors developed their findings using an analytical framework and predefined inclusion and exclusion criteria. According to the study, red meat intake had no significant impact on insulin sensitivity, fasting glucose, fasting insulin, glycated hemoglobin, pancreatic beta-cell function, or glucagon-like peptide-1. However, the authors noted that red meat intake reduced postprandial glucose levels modestly compared to meals with reduced or no red meat intake.

There is ample evidence [15] that suggests that the daily diet of adults and older people should mainly comprise vegetables, fruits, legumes, nuts, whole grains, unsaturated vegetable oils, and fish. This eating pattern, which involves limited consumption of red and processed meat, high-fat dairy products, and refined carbohydrates or sweets, has been linked to a decreased risk of all-cause mortality.

1.3 The Growing Popularity of Meat Alternatives for the Health and Environment

With growing awareness of the effects of climate change, the negative impacts on human health, and animal welfare concerns, there has been a surge in demand for meat alternatives that are enriched with nutrients but have a lower or no environmental impact. In 2020, during the COVID-19 pandemic, the demand for plant-based meat, dairy, and eggs increased substantially as more consumers cooked at home, focused on their health, and reconsidered their impact on the planet. The market for meat substitutes is experiencing significant growth because of both consumer concerns and the ability of manufacturers to offer plant-based alternatives with superior texture, flavor, and mouthfeel properties. The demand in this sector is mainly driven by Gen Z and Millennials, who are focused on health, climate change, and animal cruelty issues. Many companies are teaming up with brands to provide a wide range of products catering to various dietary needs and preferences. Even non-vegetarian consumers are embracing plant-based options due to reasons such as improved nutrition, weight management, concern for animal welfare, and environmental sustainability. Another popular trend is the use of whole vegetables and grains in ingredient lists, which appeals to health-conscious buyers seeking recognizable ingredients on product labels. Consumers also look for third-party certifications on packaging, which indicate if the product is non-GMO, gluten-free vegan, or kosher certified. Therefore, the increasing demand for plant-based alternatives along with substantial investments in innovative products are anticipated to continue creating profitable opportunities for market expansion. The global market for meat substitutes is projected to grow from $5.88 billion in 2022 to $12.30 billion by 2029, with a CAGR of 11.11% during the forecast period from 2022 to 2029 [16]. Plant-based protein segment held the largest market share of 62.45% in 2023 and is expected to maintain dominance over the forecast period. However, the mycoprotein segment is projected to register the fastest growth during the forecast period, with a CAGR of 43.5% from 2024 to 2030.

There are several alternative forms of meat available in the market today, such as plant protein, insects, mycoprotein (protein derived from fungi), and lab-grown meats [17]. These alternatives are gaining popularity among food enthusiasts and are used to produce meat analogues. Analogues are similar in structure to one another but differ slightly in composition. Meat analogue or mimic meat is defined as a foodstuff that has a similar structure to meat but is significantly different in composition [18]. These are also called as most plant-based meat, vegetarian meat, meat substitute, mimic meat, meat replacement, synthetic meat, or amalgam meat. Plant-based meats are the most common among the alternative meat categories. These meat substitutes rely on plants to create a flavorful and meat-like experience [19]. They are based on recipes that have been developed over several decades, using a variety of ingredients such as mushrooms, rice, lentils, soy protein, and wheat gluten. Meat-like flavorings are added to these ingredients to create a finished product that closely resembles meat [20]. Soy-based products like tempeh and tofu are popular meat substitutes, as well as plant proteins such as wheat gluten, which are commonly used in foods like seitan. Another type of meat substitute is made from textured vegetable protein (TVP), a dry product derived from soy concentrates. The products made from plant protein and mycoprotein possess the texture, flavor, and nutritional properties of meat but are made from non-animal sources. Target plant proteins isolated from plants are put through hydrolysis to increase their functionality. Plant proteins are combined with components such as flour, food adhesives, and plant-based oil to create a meat texture. Protein inputs for new plant-based meat are plentiful and reasonably priced. Hence, the volume of meat substitutes from plant-based protein sources is high and is the main factor for the dominance of the segment. Owing to higher content of nutrients like fiber that helps in controlling blood cholesterol and blood sugar, consumers prefer mycoprotein food [21]. Further, mycoprotein makes feel fuller than some animal proteins like chicken. This helps prevent overeating and weight gain. In addition, mycoprotein provides all essential amino acids, which is rarely found in other protein sources [22]. Quorn, a mycoprotein brand, offers super-protein, healthy meals for consumers. For vegans and vegetarians, meat alternatives were earlier limited to soya chunks, mushrooms, and cottage cheese.

The market for meat substitutes is still developing with companies holding a strong presence in this space. New entrants are launching products, while other key players are planning to introduce their own products due to increasing health concerns, as animal-borne diseases rise globally. Some of the initiatives taken so far include the following: In August 2023, Nestlé launched a range of plant-based cooking solutions under the Maggi Veg brand in Chile, featuring plant-based minced meat and lentil and vegetable soup. These products are designed to be accessible, affordable, and sustainable with a longer shelf-life than frozen or fresh products. In January 2022, KFC introduced Beyond Fried Chicken, made from plant-based meats, at various locations globally. In December 2021, ITC Ltd. launched plant-based meat products in anticipation of the burgeoning demand for meat substitutes and vegan meals in India. The key companies in the meat substitutes market include Amy's Kitchen, Beyond Meat, Impossible Foods, Quorn Foods, Kellogg's, Unilever, Meatless B.V., VBites Foods, SunFed, and Tyson Foods Inc.

In-depth analysis of Life Cycle Assessment (LCA) studies provides insights into alternative proteins' environmental benefits and is illustrated in Table 1.1 [23, 24]. The table displays the findings of publicly available and externally commissioned industry LCA studies, along with relevant peer-reviewed academic LCAs, published between January 2021 and August 2023. The studies compare alternative protein products with conventional meat products. As different studies use different methodologies, the results cannot be precisely compared. Impact reductions are calculated as follows: (impact of conventional meat product—impact of alternative protein product)/(impact of conventional meat product). Negative (−) values indicate reductions in impact within the environmental category, while positive (+) values indicate increases in impact. A dash (—) sign indicates that data is unavailable. The studies on raw/uncooked products are marked with an asterisk, while those on cooked products are marked with an asterisk and a dagger. The carbon opportunity of reduced land use is included in one study, marked with a double dagger. Other studies do not calculate the carbon opportunity of reduced land use. If the freed-up land is used to mitigate climate change, the benefits will be significantly greater than the direct impact reductions measured. The study conducted by Sinke et al. in 2023 is based on the 2030 scenarios. These scenarios make use of conservative assumptions for high-scale production of cultivated meat, coupled with low carbon footprint benchmarks for traditional meat production. The cultivated meat products were produced using renewable energy sources.

Bryant (2022) conducted a review of 43 studies to evaluate the potential of plant-based animal product alternatives in addressing concerns related to health and environmental sustainability [24]. The results indicated that these alternatives were more sustainable compared to animal products, as they have lower greenhouse gas emissions, water use, and land use. The author also noted that plant-based alternatives have several health benefits, including generally favorable nutritional profiles, promoting weight loss and muscle synthesis, and catering to specific health conditions. Moreover, the studies suggest that these alternatives can further improve their

Table 1.1 Comparative life cycle assessments

Producing this alternative protein	Instead of this conventional meat	Reduces GHG emissions by (%)	Reduces land use by (%)	Reduces air pollution (particulate matter) by (%)	Study
Impossible burger	Beef burger patty	−89	−96	–	Khan, et al. "Comparative environmental LCA of the impossible burger with conventional ground beef burger." (2019).
Beyond burger 3.0	Beef burger patty	−90	−97	–	Heller, et al. ""Beyond burger 3.0 life cycle assessment." (2023).
Beyond burger 1.0	Beef burger patty	−89	−92	–	Heller, et al. "Beyond meat's beyond burger life cycle assessment: A detailed comparison between a plant-based and an animal-based protein source." (2018).
Quorn fillet	Chicken breast	−75	−78	–	Kazer, et al. "Quorn footprint comparison report." (2021).
Quorn sausage	Pork primal cuts	−88	−79	–	Kazer, et al. "Quorn footprint comparison report." (2021).
Quorn mince	Beef mince	−96	−94	–	Kazer, et al. "Quorn footprint comparison report." (2021).
Morningstar original Chik patties	Chicken sausage patty	−46	−84	−69	Dettling, et al. "A comparative life cycle assessment of plant-based foods and meat foods." (2016).
MorningStar farms original sausage patties	Pork sausage patty	13	−47	−31	Dettling, et al. "A comparative life cycle assessment of plant-based foods and meat foods." (2016).
MorningStar farms grillers original burger	Beef burger patty	−84	−93	−92	Dettling, et al. "A comparative life cycle assessment of plant-based foods and meat foods." (2016).

(continued)

Table 1.1 (continued)

Producing this alternative protein	Instead of this conventional meat	Reduces GHG emissions by (%)	Reduces land use by (%)	Reduces air pollution (particulate matter) by (%)	Study
Plant-based burger patty made with soy protein	Beef burger patty	−98	−87	−99	Saerens, et al. "life cycle assessment of burger patties produced with extruded meat substitutes." (2021).
Plant-based burger patty made with soy protein	Chicken burger patty	−90	−82	−90	Saerens, et al. "Life cycle assessment of burger patties produced with extruded meat substitutes." (2021).
Plant-based burger patty made with soy protein	Pork burger patty	−90	−85	−90	Saerens, et al. "Life cycle assessment of burger patties produced with extruded meat substitutes." (2021).
Plant-based burger patty made with pea protein	Beef burger patty (Brazilian beef)	−77	−96	−58	Saget, et al. "Comparative life cycle assessment of plant and beef-based patties, including carbon opportunity costs." (2021).
Plant-based burger patty made with pea protein	Beef burger patty (Brazilian beef)	−86	−96	−58	Saget, et al. "Comparative life cycle assessment of plant and beef-based patties, including carbon opportunity costs." (2021).
Soy protein burger patty	Beef burger patty	−82	−84	−95	Smetana, et al. "Meat substitution in burgers: Nutritional scoring, sensorial testing, and life cycle assessment." (2021).
Pea protein burger patty (beyond meat)	Beef burger patty	−84	−64	−91	Smetana, et al. "Meat substitution in burgers: Nutritional scoring, sensorial testing, and life cycle assessment." (2021).

(continued)

Table 1.1 (continued)

Producing this alternative protein	Instead of this conventional meat	Reduces GHG emissions by (%)	Reduces land use by (%)	Reduces air pollution (particulate matter) by (%)	Study
Mycoprotein burger patty (Quorn)	Beef burger patty	−82	−69	−91	Smetana, et al. "Meat substitution in burgers: Nutritional scoring, sensorial testing, and life cycle assessment." (2021).
Pea protein meatball	Beef meatball (Brazilian beef)	−89	−97	−82	Saget, et al. "Substitution of beef with pea protein reduces the environmental footprint of meat balls whilst supporting health and climate stabilisation goals." (2021).
Pea protein meatball	Beef meatball (Brazilian beef)	−93	−97	−82	Saget, et al. "Substitution of beef with pea protein reduces the environmental footprint of meat balls whilst supporting health and climate stabilisation goals." (2021).
Soy-based schnitzel (vegan)	Pork schnitzel	−46	−45	−74	Mierlo, et al. "Moving from pork to soy-based meat substitutes: Evaluating environmental impacts in relation to nutritional values." (2022).
Cultivated beef	Conventional beef (from non-dairy cattle)	−92	−90	−94	Sinke, et al. "Ex-ante life cycle assessment of commercial-scale cultivated meat production in 2030." (2023). Δ

(continued)

Table 1.1 (continued)

Producing this alternative protein	Instead of this conventional meat	Reduces GHG emissions by (%)	Reduces land use by (%)	Reduces air pollution (particulate matter) by (%)	Study
Cultivated chicken	Conventional chicken	3	−64	−20	Sinke, et al. "Ex-ante life cycle assessment of commercial-scale cultivated meat production in 2030." (2023). Δ
Cultivated pork	Conventional pork	−44	−67	−42	Sinke, et al. "Ex-ante life cycle assessment of commercial-scale cultivated meat production in 2030." (2023). Δ
Hybrid cultivated meat and plant-based burger patty (blended with soy protein isolate)	Beef burger patty (produced in USA)	−87	−90	–	Kim, et al. "Environmental life cycle assessment of a novel cultivated meat burger Patty in the United States." (2022).

healthiness by using optimal ingredients and processing. The first study that evaluated the nutritional quality and ingredients of Brazilian meat substitutes found that vegan meat substitutes had similar energy and protein values compared to meat-based products, with only a few exceptions among samples, such as vegan canned fish alternatives containing less protein [25]. Overall, vegan products did not differ in terms of sodium levels, but they contained higher amounts of nutrients required for a lunch or dinner meal. Vegan meat substitutes contained higher carbohydrates and dietary fiber concentrations, and there were only a few differences regarding total and saturated fat. The authors concluded that vegan meat substitutes may help people adhere to and maintain vegan and vegetarian diets.

Plant-based alternatives to meat products are also criticized for being highly processed, with their texture, flavor, and appearance mimicking those of conventional meat [26]. For instance, plant-based burgers, sausages, and minced "meats" are designed to change color from raw red pink to cooked brown, which is achieved by using colorants like beetroot powder and reducing sugars such as dextrose, xylose, or arabinose that react when heated. Many additives are also used to create a meat-like bite and juicy texture, to improve slicing ability and texture when served cold, and to adjust the taste and texture of plant protein to resemble animal protein. These additives include methylcellulose, carrageenan, monosodium glutamate, soy lecithin, stabilizers, and fillers, which can have varied effects on the body. From a nutritional standpoint, these additives, fortifications, and manipulations mean that a high proportion of these products fall into the category of ultra-processed foods, which are typically linked to poor health outcomes. It is unclear if alternative meats significantly reduce carbon footprint. This is because the process of preparing alternative meats requires a lot of resources. For example, "chicken-free" Quorn contributes around 3 kg CO_2 equivalent per kg, while other mycoproteins can be up to 6 kg CO_2 equivalent per kg, which is the same as poultry production [27]. Current estimates suggest that in vitro meat would have a higher carbon footprint than poultry or pork. Although the environmental impact is cited as a significant driver of consumer motivation for alternative proteins, research suggests that consumers are often unaware of the environmental impact of food production and the relatively larger contribution of other lifestyle behaviors, such as international flights, to GHG emissions.

1.4 The Importance of Scientific Research in Understanding the Impact of Meat Substitutes on Global Health Outcomes

In recent years, alternative meats such as plant-based and cell-cultured options have gained significant attention as potential solutions to address concerns associated with conventional meat production, including environmental impact, ethical issues, and health risks. While these alternatives offer benefits such as reduced greenhouse gas emissions, improved animal welfare, and decreased reliance on land and water resources, further research is essential to comprehensively evaluate their safety, impact on food security, and nutritional quality.

One of the foremost concerns surrounding alternative meats is their safety for human consumption. While plant-based alternatives are generally considered safe, with rigorous regulatory oversight ensuring compliance with food safety standards, cell-cultured meats present unique challenges. Research is needed to assess the safety of cell-cultured meat production processes, including cell sourcing, growth media composition, and potential microbial contamination risks. Comprehensive studies are required to evaluate the presence of pathogens, allergens, and other contaminants in cell-cultured meats and to establish robust safety protocols to mitigate any risks to consumer health.

Adoption of alternative meat must be accompanied by research to assess scalability, affordability, and accessibility. Studies are needed to evaluate the feasibility of large-scale production of plant-based and cell-cultured meats, including the optimization of production processes, resource efficiency, and economic viability. Additionally, research should explore the potential impact of alternative meat production on agricultural land use, water usage, and biodiversity conservation. This is to ensure that these technologies contribute positively to long-term food security goals. It is imperative to understand the nutritional value of alternative meats to make informed dietary decisions. Although plant-based meats strive to imitate the taste and texture of traditional meats, quality research is needed to determine their nutritional adequacy and potential health benefits or risks. To compare alternative and traditional meat products, comparative randomized control trial studies should evaluate macronutrients and micronutrients, including protein quality, fat composition, and vitamin and mineral content. Similarly, cell-cultured meats require research on their nutritional content, including protein, fat, and essential nutrient profiles, to determine their suitability as a dietary source of protein.

Advancing research on alternative meats is essential to address critical questions related to safety, food security, and nutritional quality. By conducting rigorous scientific studies, we can ensure that alternative meat products meet stringent food safety standards, contribute to global food security goals, and provide consumers with nutritious and sustainable dietary choices. Collaborative efforts between academia, industry, and regulatory agencies are needed to prioritize research in this field. This will accelerate the development and adoption of safe, nutritious, and environmentally friendly alternative meat options. Only through evidence-based research can we unlock the full potential of alternative meats to reshape food production and consumption.

References

1. Stanford Woods Institute for the Environment. Meat's Environmental Impact. Retrieved February 13, 2024, from https://woods.stanford.edu/news/meats-environmental-impact.
2. Global meat industry – statistics & facts | Statista. Retrieved February 13, 2024, from https://www.statista.com/topics/4880/global-meat-industry/#topicOverview.
3. Ferreira JP, Sharma A, Zannad F. The future of meat: health impact assessment with randomized evidence. Am J Med. 2021;134(5):569–75. https://doi.org/10.1016/j.amjmed.2020.11.007.
4. Clark M, Tilman D. Comparative analysis of environmental impacts of agricultural production systems, agricultural input efficiency, and food choice. Environ Res Lett. 2017;12(6):064016. https://doi.org/10.1088/1748-9326/AA6CD5.
5. Beef | Industries | WWF. Retrieved February 13, 2024, from https://www.worldwildlife.org/industries/beef.
6. Poore J, Nemecek T. Reducing food's environmental impacts through producers and consumers. Science. 2018;360(6392):987–92. https://doi.org/10.1126/science.aaq0216. Erratum in: Science. 2019 Feb 22;363(6429)
7. Rowe SJ, Hickey SM, Bain WE, Greer GJ, Johnson PL, Elmes S, et al. Can we have our steak and eat it: the impact of breeding for lowered environmental impact on yield and meat quality in sheep. Front Genet. 2022;13:911355. https://doi.org/10.3389/fgene.2022.911355.
8. Kirwan LB, Walton J, Flynn A, Nugent AP, Kearney J, Holden NM, et al. Assessment of the environmental impact of food consumption in Ireland-informing a transition to sustainable diets. Nutrients. 2023;15(4):981. https://doi.org/10.3390/nu15040981.
9. Clark M, Springmann M, Rayner M, Scarborough P, Hill J, Tilman D, et al. Estimating the environmental impacts of 57,000 food products. Proc Natl Acad Sci USA. 2022;119(33):e2120584119. https://doi.org/10.1073/pnas.2120584119.
10. Sranacharoenpong K, Soret S, Harwatt H, Wien M, Sabaté J. The environmental cost of protein food choices. Public Health Nutr. 2015;18(11):2067–73. https://doi.org/10.1017/S1368980014002377.
11. Yan D, Liu K, Li F, Shi D, Wei L, Zhang J, et al. Global burden of ischemic heart disease associated with high red and processed meat consumption: an analysis of 204 countries and terri-

tories between 1990 and 2019. BMC Public Health. 2023;23(1):2267. https://doi.org/10.1186/s12889-023-16954-4.
12. Gao J, Guo X, Wei W, Li R, Hu K, Liu X, et al. The Association of Fried Meat Consumption with the gut microbiota and Fecal metabolites and its impact on glucose homoeostasis, intestinal endotoxin levels, and systemic inflammation: a randomized controlled-feeding trial. Diabetes Care. 2021;44(9):1970–9. https://doi.org/10.2337/dc21-0099.
13. Landry MJ, Ward CP, Cunanan KM, Durand LR, Perelman D, Robinson JL, et al. Cardiometabolic effects of omnivorous vs vegan diets in identical twins: a randomized clinical trial. JAMA Netw Open. 2023;6(11):e2344457. doi: https://doi.org/10.1001/jamanetworkopen.2023.44457. Erratum in: JAMA Netw Open. 2023;6(12):e2350422.
14. Sanders LM, Wilcox ML, Maki KC. Red meat consumption and risk factors for type 2 diabetes: a systematic review and meta-analysis of randomized controlled trials. Eur J Clin Nutr. 2023;77(2):156–65. https://doi.org/10.1038/s41430-022-01150-1.
15. English LK, Ard JD, Bailey RL, Bates M, Bazzano LA, Boushey CJ, et al. Evaluation of dietary patterns and all-cause mortality: a systematic review. JAMA Netw Open. 2021;4(8):e2122277. https://doi.org/10.1001/jamanetworkopen.2021.22277.
16. Meat Substitutes Market Size, Growth, Potential, Report, 2031. Retrieved February 13, 2024, from https://www.fortunebusinessinsights.com/industry-reports/meat-substitutes-market-100239.
17. Ahmad M, Qureshi S, Akbar MH, Siddiqui S, Gani A, Mushtaq M, Hassan I, et al. Plant-based meat alternatives: compositional analysis, current development and challenges. Appl Food Res. 2022;2(2):100154. https://doi.org/10.1016/j.afres.2022.100154.
18. Kołodziejczak K, Onopiuk A, Szpicer A, Poltorak A. Meat analogues in the perspective of recent scientific research: a review. Foods. 2021;11(1):105. https://doi.org/10.3390/foods11010105.
19. Hoffman R. Defining "plant-based meats". Eur J Clin Nutr. 2024;78(1):72–3. https://doi.org/10.1038/s41430-023-01340-5.
20. Wang Y, Tuccillo F, Lampi AM, Knaapila A, Pulkkinen M, Kariluoto S, et al. Flavor challenges in extruded plant-based meat alternatives: a review. Compr Rev Food Sci Food Saf. 2022;21(3):2898–929. https://doi.org/10.1111/1541-4337.12964.
21. Rubio NR, Xiang N, Kaplan DL. Plant-based and cell-based approaches to meat production. Nat Commun. 2020;11(1):6276. https://doi.org/10.1038/s41467-020-20061-y.
22. Souza Filho PF, Andersson D, Ferreira JA, Taherzadeh MJ. Mycoprotein: environmental impact and health aspects. World J Microbiol Biotechnol. 2019;35(10):147. https://doi.org/10.1007/s11274-019-2723-9.
23. Hayek MN, Harwatt H, Ripple WJ, Mueller ND. The carbon opportunity cost of animal-sourced food production on land. Nat Sustain. 2020;4(1):21–4. https://doi.org/10.1038/s41893-020-00603-4.
24. Bryant CJ. Plant-based animal product alternatives are healthier and more environmentally sustainable than animal products. J Future Foods. 2020;6:100174. https://doi.org/10.1016/j.fufo.2022.100174.
25. Romão B, Botelho RBA, Nakano EY, Raposo A, Han H, Vega-Muñoz A, et al. Are vegan alternatives to meat products healthy? A study on nutrients and Main ingredients of products commercialized in Brazil. Front Public Health. 2022;10:900598. https://doi.org/10.3389/fpubh.2022.900598.
26. Suri S, Ray S. Will a plant-based approach be sufficient to address food security, nutritional adequacy and sustainability in the future? Environ Urban ASIA. 2023;14(2):299–308. https://doi.org/10.1177/09754253231193130.
27. Tso R, Lim AJ, Forde CG. A critical appraisal of the evidence supporting consumer motivations for alternative proteins. Foods. 2021;10:24. https://doi.org/10.3390/foods10010024.

Chapter 2
Understanding Meat Substitutes

2.1 Definition and Classification of Meat Substitutes

Meat alternatives are extensively researched since past few years, but the terminology around meat alternatives is still unclear. According to the published literature [1], meat alternatives are referred to as meat replacements, fake meat, faux meat, mock meat, meat substitutes, and meat analogues. Many perceive "plant-based meats" as the only meat alternative available in the consumer market. Evidence is poor on the similarity of these terminologies. Researchers, food scientists, food-producing companies, and other stakeholders often struggle to evaluate the drawbacks of meat alternatives when there are no clear criteria that define them as meat alternatives. Simultaneously, any research on meat alternatives can be contested in the absence of a definition. The common notion suggests that meat substitutes are a range of products that imitate the taste, texture, and nutritional value of traditional meat while being more sustainable [2]. There are several types of meat substitutes, including tofu-based, tempeh-based, TVP-based, seitan-based, and Quorn-based alternatives [1]. In addition to these, there are emerging categories such as cultured meat, which is grown in labs and derived from stem cells to closely resemble conventional meat. Plant-based substitutes, also called meat analogues or imitation meat, are made from plant-based ingredients to replicate the taste, texture, and appearance of meat. Other alternative sources, such as insect-based, fungi-based,

The original version of the chapter has been revised. A correction to this chapter can be found at https://doi.org/10.1007/978-3-031-94604-2_11

S. Ray, S. Suri, *The Sustainable Plate: Unveiling the Science of Meat Substitutes and Impact on Global Health*, https://doi.org/10.1007/978-3-031-94604-2_2

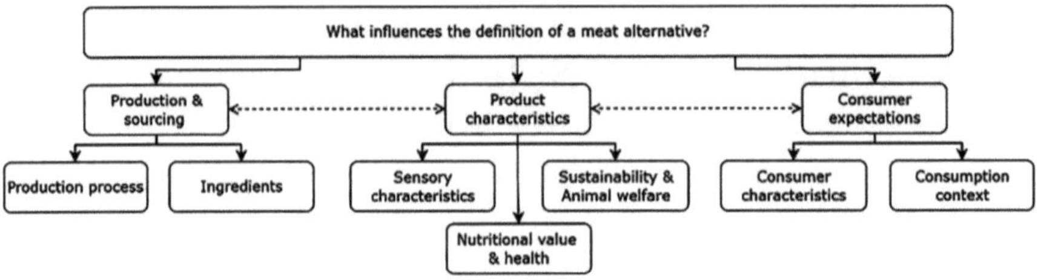

Fig. 2.1 The drivers in defining meat alternatives. (Adopted from Ketelings et al. [1])

and single-cell-based substitutes, are gaining recognition for their potential to provide sustainable protein options [1]. This classification system helps to understand the diverse landscape of meat substitutes, catering to a variety of dietary preferences and sustainability considerations.

A scoping review [1] of evidence between 2000 and 2021 by Ketelings et al. (2023) aimed to define and classify meat alternatives by assessing 193 eligible literatures. Based on the analysis, the authors have developed a taxonomy (Fig. 2.1) that describes the dimensions that define a meat alternative. According to the authors, there are three facets that influence what is considered a meat alternative. The product characteristics displayed in the figure are influenced by the other two facets: production and sourcing and consumer expectations. However, these two facets also affect how these product characteristics are formed.

The first facet, i.e., production and sourcing, significantly influence their characteristics, encompassing taste, appearance, texture, and nutritional composition [3]. Numerous ingredients are utilized and selected for their sensory attributes, functional properties, and nutrient content to develop these alternatives. To compensate for protein content in meat sources, common protein sources used in these products include soy, mycoprotein, lentils, and pulses, with soy protein being widely utilized due to its favorable functional properties and balanced amino acid composition [4]. Additionally, novel plant-based proteins like seaweed, algae, and insect-derived proteins are gaining recognition for their nutritional benefits and sustainability [5]. The inclusion of animal-based ingredients such as egg, milk, or cheese is also observed in some meat alternatives. Notably, cultured meat, produced through tissue engineering techniques, represents a cutting-edge development in the field, blurring the line between traditional meat and meat alternatives [6]. The composition of meat alternatives typically involves protein, constituting approximately 20–50% of the product, supplemented with lipids, polysaccharides, flavorings, colorants, and fortifying ingredients to enhance texture, taste, and nutritional value [1]. The production process encompasses varying degrees of ingredient processing, ranging from unprocessed primary agricultural ingredients like pulses and nuts to highly processed forms such as textured vegetable proteins and mycoproteins such as Quorn [5]. While some alternatives, like tofu, may require minimal processing, others aim to closely mimic the sensory experience of meat through extensive processing. This ultra-processed nature raises concerns regarding nutrient density and overall health implications, prompting discussions about the potential trade-offs between sensory appeal and nutritional quality [7]. Despite these considerations, ongoing research and development aim to refine meat alternatives, enhancing their sensory characteristics and nutritional profiles to offer consumers healthier and more sustainable dietary options.

The sensory characteristics that influence the definition of meat alternatives include texture, flavor, appearance, smell, and taste. Some scholars believe that the acceptance of meat alternatives by most consumers is driven by what closely resembles conventional meat products [8]. However, some recognize the diversity within the category of meat alternatives to be acceptable to consumers [9]. They acknowledge options that may not necessarily aim to replicate meat but still offer viable alternatives, such as tofu and tempeh. Nutritional value and health considerations are crucial in evaluating the efficacy of meat alternatives. While some people advocate for meat substitutes to match the sensory and nutritional attributes of meat products, empirical evidence suggests that achieving this parity can be challenging [10]. According to the author, meat alternatives often contain higher levels of sodium and sugar and may lack essential micronutrients like vitamin B12 and iron. Despite efforts to fortify these products, disparities in nutrient composition persist, raising concerns about their ability to provide adequate nutrition compared to conventional meat. Additionally, the health perception of meat alternatives remains divided. Debates over their potential benefits in reducing antibiotic use and mitigating foodborne illnesses are countered by concerns over their highly processed nature and uncertain overall health impact. Sustainability is a compelling argument for the adoption of meat alternatives. Reducing meat consumption can alleviate the environmental pressures associated with livestock production [11]. While meat substitutes are generally perceived as more sustainable due to their reduced reliance on animal agriculture, questions remain regarding the environmental footprint of their production processes. The classification of meat alternatives as ultra-processed foods raises concerns about the additional energy and resources required, potentially offsetting their sustainability benefits [12]. Despite consumer motivations to choose meat alternatives for environmental reasons, the lack of standardized sustainability labelling and information hampers informed decision-making. This highlights the need for comprehensive standards and guidelines to assess the sustainability of meat alternative products.

The labelling and marketing strategies for meat alternatives depend on how they are defined, which influences consumer expectations and target demographics [1]. For instance, using labels like "burger" or "sausage" implies that the product resembles meat and creates specific sensory expectations for consumers. Although there is a general agreement on the sensory aspects, there is still uncertainty regarding the nutritional values. Labelling products as meat-like may make consumers expect similar nutritional profiles, leading to potential confusion. Moreover, using terms like "tofu" and "vegan" may elicit negative responses from omnivorous consumers, while "plant-based" is more widely accepted. Companies adjust the labelling to target specific consumer groups, as preferences vary based on dietary choices and lifestyle.

The use of meat-like terminology on labels and in marketing for meat alternatives has sparked debate both in the European Union (EU) and the United States [13]. In the EU, a proposal to reserve terms like "burger," "sausage," and "steak" exclusively for products containing meat was drafted in 2019 but ultimately rejected by the European Parliament in October 2020 [14]. This decision allows the continued use of meat-related terminology on meat alternatives. While some argue that such terminology could be misleading, others contend that it helps create consumer expectations and is not inherently deceptive. Similarly, in the United States, lawmakers in several states have passed statutes prohibiting the use of "meat" in the labelling of plant-based meat products, citing concerns about consumer confusion and public health [1]. However, meat alternative companies argue that such regulations infringe upon their rights to free speech and impede fair competition in the market. In contrast, the Plant Based Food Association introduced voluntary labelling standards in December 2019, allowing references to animal-based terminology on meat alternative products if the label clearly indicates their plant-based or vegetarian nature [15]. Despite ongoing political debates and regulatory actions, questions remain regarding the impact of labelling and marketing meat alternatives with meat terminology on consumer behavior. More research is needed to understand whether such labelling practices are confusing or misleading for consumers. Resolving this debate requires a nuanced examination of consumer perceptions and behaviors in response to different labelling strategies, ultimately informing future regulatory decisions and industry practices surrounding alternative meat labelling [16].

The quest for the perfect meat alternative remains elusive in today's market, as no single product can fully satisfy the diverse expectations of different consumer segments. The largest group of consumers seeking meat alternatives are omnivores looking to reduce meat consumption without sacrificing the sensory experience of eating meat [17]. For this segment, characteristics such as taste, texture, and appearance are paramount, along with nutritional considerations to match or exceed the benefits of meat. However, smaller segments may prioritize factors like novelty, sustainability, or health benefits over meat-like attributes. Manufacturers must tailor products to meet the specific expectations of their target consumers, balancing sensory appeal, nutritional value, and ethical considerations. The definition is still confusing and missing. However, the scoping review in 2023 provides a promising framework for developing a solid taxonomy that can help academics define meat alternatives [1].

2.2 Various Types of Meat Alternatives and Their Global Producers

Meat alternatives refer to products devoid of animal ingredients, aiming to replicate the sensory experience typically associated with meat. This category has witnessed a surge in popularity across developed nations, attributed to advancements in technology and the introduction of novel ingredients capable of mimicking the taste, texture, and functionality of traditional meat products. The meat alternatives are available in various forms, including burgers, sausages, patties, nuggets, and crumbles, catering not only to vegans and vegetarians but also to a broader demographic of meat reducers, commonly referred to as flexitarians, as well as meat eaters. This expanded appeal underscores the versatility and accessibility of meat alternatives in accommodating diverse dietary preferences and lifestyles. With the increasing popularity of meat alternatives, there are now various types of meat alternatives available for consumption.

Meat alternatives refer to a wide variety of options, including plant-based alternatives like soy, pea, and gluten, as well as emerging technologies like cell-based and fermentation-based alternatives [18]. Recent advancements in this field have led to the introduction of new protein sources, such as microalgae proteins derived from Spirulina and proteins isolated from insects. Cell-based meat, which is cultivated through in vitro techniques, and mycoproteins produced via fungal fermentation offer unique advantages such as the ability to form elongated fibers that impart a meat-like chewable texture and provide potential health benefits [19]. However, the high production costs, projected climate impacts, and other factors have hindered the widespread adoption of these alternative proteins. Despite ongoing development and efforts to improve these technologies, they remain in the early stages of commercialization. Major meat alternatives are discussed in the next segment.

2.2.1 Plant-Based Meat Alternatives

The idea of meat alternatives as a source of protein has been around for centuries, with traditional products like tempeh, tofu, and Seitan being some of the most popular examples. Tofu, which is considered a typical meat analogue, originated in China during the Han dynasty (206 BC–220 AD) and became widely popular during the Tang dynasty (618–907) [20]. Its popularity likely spread to Japan during the later Tang or early Song dynasty. In the early twentieth century, pioneers like John Harvey Kellogg introduced nut and cereal-based products such as Nuttose and Protose, with the aim of promoting good health. Along with traditional Asian products, the concept of dry texturized vegetable protein emerged, derived from extruded defatted soy meal, soy protein concentrates, or wheat gluten. After World War II, advancements in the production and packaging industry led to significant progress in plant protein concentrates, isolates, and textured proteins. These developments facilitated the creation of soy-based meat alternatives, especially during periods of surging meat consumption in industrialized nations. In the USA, products like Tofurky, which cater to a niche vegetarian demographic, emerged in 1980. Burger King introduced the first American meat alternative to enter the food chain in 2002 with a traditional plant-based burger. In the new millennium, growing awareness about health and sustainability fueled demand for alternatives to conventional meat. This led to the emergence of new products like Impossible Burger and Beyond Burger, marking a new generation of analogues and doubling the plant-based meat industry. With the help of advancements in food science and manufacturing, plant-based meat aims to recreate the taste, texture, appearance, and functionality of conventional meat products like sausages, burgers, and fillets. These meat alternatives are made from combinations of plant proteins, fats, gums, and spices and using extruders or unique processing technologies to achieve robust consumer acceptability globally.

Meat alternatives often use soy protein as it is cost-effective and has desirable characteristics. Other oil-seed plants and proteins produced by microorganisms on various substrates have also been used in making meat replacements. Currently, proteins from cereals like rice, wheat, and maize, as well as defatted oil seed and bean flour, cereal derivatives, and defatted soy flour and soy protein concentrates, are also used to make meat alternatives. Soy meat, also known as textured vegetable protein, is derived from soybeans, primarily in Asian countries. The production process involves hot extrusion of defatted soy proteins, resulting in expanded high-protein chunks, nuggets, strips, grains, and other shapes. These denatured proteins give the textured vegetable protein a fibrous consistency similar to meat, making it versatile for various culinary applications. Textured vegetable protein can absorb water or other liquids multiple times its weight, making it suitable for use as a meat extender or as a standalone meat alternative. Unlike meat extenders, which do not resemble meat when cooked alone, meat analogues made from textured vegetable protein closely resemble meat in appearance, color, flavor, and texture when hydrated and cooked. Tofu, another popular plant-based meat alternative, is a traditional Asian food made from soy protein. It is widely recognized for its outstanding protein, calcium, and iron content. Tofu, also known as bean curd, is popular in various countries and is highly versatile in cooking. It absorbs flavors well and can be prepared in various ways, making it a staple in many cuisines. Tofu is also renowned for its unique nutritional profile, being low in cholesterol and calories while rich in protein, vitamins (such as vitamins A, C, D, E, and various B vitamins), minerals (including calcium, phosphorus, potassium, magnesium, iron, zinc, and selenium), and omega-3 fatty acids [21]. One block of hard tofu contains essential nutrients like calcium, phosphorus, potassium, magnesium, iron, zinc, and folate, making it a nutritious addition to any diet. Tempeh, originating from Indonesia, is another meat alternative made from fermented soybeans and cooked soybeans, often mixed with rice or millet, and inoculated with the Rhizopus oligosporus culture [20]. Unlike tofu, tempeh retains the whole soybeans, giving it a thicker, "meatier" consistency. It is rich in protein, with an average protein content of about 19.5%, comparable to chicken, beef, eggs, and milk. The fermentation process not only enhances the protein content but also increases the bioavailability of nutrients, making tempeh a highly nutritious food. Additionally, tempeh is a good source of fiber and essential nutrients, including Vitamin B12, which is produced as a by-product of the fermentation process.

Seitan, which is also known as wheat meat or wheat gluten, is a popular meat substitute for vegetarians [20]. It is made by continuously washing wheat flour dough until wheat-based protein gluten is obtained. Seitan is a great alternative for meat lovers who are not sensitive to gluten, but individuals who have gluten sensitivities tend to avoid it. Seitan is a cost-effective option as it is made from simple raw materials and can be prepared in various ways. It is also versatile and can be seasoned according to taste. Since wheat crops are cultivated worldwide, Seitan has a great potential for global production. The consistency of Seitan closely resembles the chewy fibers found in meat, making it a desirable option for vegetarian meat production. Sweet lupines are also used in the production of vegetarian meat alternatives. A Dutch company, Meatless B.V., has successfully developed 100% vegetable fibers from lupine or wheat. These fibers come in various shapes, colors, and flavors, offering versatility in meat replacement. Meatless fibers can replace a significant portion of meat and are primarily used to create hybrid products and meat alternatives. Although other lupine-based meat alternatives have been developed recently, they have yet to gain significant traction in the market.

Risofu is a range of rice-based burgers and sausages developed by Bahama Rice Burger, a company based in the United States [20]. The term "Risofu" comes from the Italian word "riso," which means rice, and "tofu," which refers to rice tofu. The idea for Risofu was inspired by the Shan region in Thailand, where rice-based tofu is commonly produced. To make Risofu, the company combines different varieties of white, brown, and wild rice to maximize the nutritional value of the final product. Apart from rice-based alternatives, algae is emerging as a potential source of vegetarian meat substitutes. Algae includes seaweed and microalgae, which provide a sustainable and innovative protein source. Seaweed is a multicellular organism, while microalgae includes both single-celled and multicellular forms like Chlorella and diatoms. Meat analogues made from microalgae can be grown in various environments and have a lower environmental impact than conventional meat sources like beef, pork, and chicken. The use of algae in meat substitutes aligns with the growing demand for sustainable and eco-friendly food production practices.

Several companies are leading the charge in producing meat substitutes, offering alternatives to traditional animal based protein products. Beyond Meat, based in El Segundo, California, specializes in pea protein-based products designed to mimic the taste and texture of meat. Another notable player is Boca Foods, a subsidiary of Kraft Foods, which produces Boca Burgers, veggie burgers made from soy protein and wheat gluten. Fry Group Foods, founded in 1991 by Wally and Debbie Fry, offers a range of vegan meat substitutes. Gardein, developed by Canadian entrepreneur Yves Potvin, utilizes soy, wheat, grains, and vegetables, including pea protein, to create its meat-free foods. Impossible Foods is renowned for its plant-based meat substitutes, which utilize genetically engineered heme from plants. LightLife is another major producer of vegetarian and vegan meat substitutes. Prime Roots, headquartered in California, specializes in koji-based meat substitutes. Quorn, a British company owned by Monde Nissin Corporation, offers a variety of meat substitute products. Simulate, based in the United States, focuses on producing plant-based chicken nuggets. Turtle Island Foods is known for its Tofurky products, providing vegetarian and vegan alternatives to turkey and other meatless options. Finally, Upside Foods, located in the Bay Area, California, produces a range of meat substitutes, contributing to the growing market for plant-based protein alternatives.

2.2.2 Fungi-Based or Mycelium Meat Alternatives

Mycoprotein, derived from fungal biomass, represents a promising protein source notable for its appealing taste, high protein content, and eco-friendly characteristics [22]. Research indicates that mycoprotein shares similar proteins with eggs and boasts a high polyunsaturated fatty acid content (40%), along with low levels of saturated fatty acids (11%) and overall fat content (3.25%). Notably, mycoprotein exhibits superior sensory properties compared to other meat alternatives, as evidenced by its high acceptability scores, which could be attributed to its fiber bundle content closely resembling meat fibers.

Better protein source than most meat alternatives: *Most animal protein advocates argue that most plants lack complete protein profiles, meaning they don't contain all nine essential amino acids our bodies require but cannot produce independently. Additionally, plant proteins can be challenging for our digestive systems to process, leading to lower bioavailability than animal proteins. However, fungi-based meat solves this issue commonly associated with plant-derived alternatives. Fungi-based proteins offer a complete amino acid profile with high bioavailability, making them a compelling substitute for beef or chicken, especially for individuals concerned with meeting their protein requirements. Beyond its protein content, unprocessed fungi products are naturally low in fat and sodium, attributes commonly linked to heart disease, diabetes, and obesity—conditions often associated with high intake of animal protein, particularly beef. Many meat alternatives on the market are heavily processed and laden with artificial additives and seed oils, whereas mycoprotein offers a wholesome alternative that may contribute positively to health and longevity. Mycelium, the underground network of fungi, is abundant in essential vitamins, minerals, and other beneficial compounds crucial for maintaining health and optimal bodily functions. Notably, mycoprotein is rich in vitamin D, a nutrient notoriously challenging to obtain from food sources, particularly for individuals with limited access to seafood or inadequate sunlight exposure. Mushrooms and mycelium offer a valuable source of this vital vitamin, along with other nutrients often lacking in plant-based diets, such as vitamin B-12, B-9, selenium, iron, copper, and zinc, making mycoprotein a valuable addition to the diets of vegans and vegetarians. One of the key factors that set fungi-based meat apart from other plant-based substitutes is its meaty texture. While many plant-based alternatives struggle to replicate the fibrous texture of meat and often require complex processing and artificial additives, fungi-based meat naturally possesses a fibrous texture akin to meat. The intricate network of hyphae, filamentous structures comprising mycelium, closely mimics the texture of meat, offering a satisfying alternative to traditional animal-based proteins. Furthermore, different fungal species yield*

(continued)

varying textures, providing versatility in culinary applications and enabling the creation of fungal substitutes for a wide range of meat cuts. Moreover, mycoprotein production demands less land and water while emitting lower greenhouse emissions compared to beef production, underscoring its environmental sustainability. With its availability in 17 countries and a projected increase in demand, mycoprotein holds significant promise as a viable alternative protein source poised for further growth in the future.

The development of mycoprotein dates back to the 1980s, when the FDA officially approved a fungi-based protein for consumption in 2002. With the growing popularity of veganism and increasing environmental awareness, there is optimism among environmentalists that fungi-based meat alternatives will gain widespread acceptance and adoption in the market. While there are not as many fungi-based meat alternatives in the market as there are soy- or pea-based options, there are several companies that specialize in producing them. Meati, a startup, offers a diverse range of products, including the Meati Classic Cutlet, Meati Crispy Cutlet, and Meati Steak Filet, among others, which are already available or soon to be launched on the market. Prime Roots, based in California, uses Koji, a Japanese fungus, to create a variety of meat substitutes such as salmon burgers, shrimp, lobster, tuna, chicken, sausage, and beef. Prime Roots claims to authentically replicate the texture and flavor of traditional deli meats. Unlike typical deli slices, Prime Roots offers whole joints, expanding its product lineup to include deli meats, charcuterie, and bacon. Their assortment of whole cuts encompasses turkey varieties like cracked pepper and smoked, as well as ham selections including smoked, black forest, and sugar shack maple. The charcuterie range features an array of options such as salami, pepperoni, pizza style pepperoni, apple-sage, black truffle patês, and foie gras torchon, while their bacon options are hickory-smoked. In 2021, Prime Roots introduced a line of plant-based ravioli, including what they claimed to be the world's first vegan lobster ravioli. Earlier this year, the company secured a minor investment from mycoprotein leader Quorn to enhance both brands' market presence and product offerings. Previously known as Atlast Food Co in 2020, MyForest Foods, a spinoff of mycelium pioneer Ecovative Design, produces a whole-cut bacon alternative from mycelium, branded as MyBacon. With minimal ingredients apart from mycelium, including salt, coconut oil, sugar, natural flavors, and beet juice for color, this vegan bacon offers a sustainable option. Additionally, Catalan brand Libre Foods, launched in 2020, utilizes precision fermentation to craft fibers resembling animal muscle fibers, starting with their bacon product. While bacon marks their initial offering, Libre aims to develop a complete range of whole cuts, including poultry and seafood alternatives. Quorn is the best-known player in this space. The brand has gained significant traction in the US market and is supplying its products to restaurants seeking eco-friendly and delicious plant-based alternatives.

2.2.3 *Insect-Based Meat Alternatives*

Insects have been part of human diets since early evolution. However, in 2013, the Food and Agriculture Organization of the United Nations (FAO) emphasized the importance of modern food science practices to promote insect trade, consumption, and acceptance [23]. This initiative gained more momentum with Regulation 2015/2283 of the European Parliament and the Council of the European Union, which classified whole insects and their parts as novel foods. The European Food Safety Authority (EFSA) provided a scientific opinion on insect consumption in 2015, identifying a list of insect species that have high potential for use in animal feed and human consumption [25]. In 2021, the EFSA issued a favorable opinion on the safety of dried yellow mealworm (Tenebrio mellitus) larvae (TM larvae), Locusta migratoria (LM), and Acheta domesticus (AD) as novel foods according to European Union Regulation 2015/2283 [25]. Many organizations have been actively studying the feasibility of using insects as a source of food and feed. The Food and Agricultural Organization of the United Nations (FAO) has conducted extensive research into the advantages of incorporating insects into human diets, considering cultural, economic, safety, production, and nutritional factors. Insect proteins are found to meet all the amino acid composition requirements outlined by the World Health Organization (WHO), with high levels of essential amino acids such as phenylalanine, tyrosine, tryptophan, lysine, and threonine. Among the different orders of insects, Orthoptera is considered one of the most valuable alternative protein sources, with many edible insects in this order containing the essential amino acids necessary for human consumption.

Insects, classified as invertebrate animals within the Phylum Arthropoda, boast remarkable nutritional value and diverse functional properties. Despite representing a staggering 95% of the animal kingdom's biomass and offering significant biodiversity, only limited studies have explored their potential applications in food systems. However, recent research indicates that insects are a promising protein source, rivalling the nutritional quality provided by traditional sources like cattle. Edible insects present a promising alternative that offers numerous health benefits while also being sustainable. Edible insects are packed with essential nutrients such as vitamin B12, iron, zinc, fiber, essential amino acids, omega-3 and omega-6 fatty acids, and antioxidants. Incorporating edible insects like crickets into the human diet could yield significant environmental and nutritional advantages. These include reducing greenhouse gas emissions, minimizing agricultural land and water use, and promoting better management of chronic diseases like diabetes, cancer, and cardiovascular disease. The environmental advantage is partly attributed to the highly efficient feed-to-meat conversion ratio of insects, with crickets, for instance, requiring only 2.2 kg of feed to produce 1 kg of edible insects. Beyond their nutritional benefits, insect proteins showcase a wide array of functional properties, including foamability, emulsifying capabilities, and gelling abilities. These multifaceted attributes position insects as a promising and sustainable protein source with diverse applications in the food industry.

Table 2.1 Nutritional composition of different edible insect species

Edible insects	Protein (%)	Fat (%)	Fiber (%)	NFE (%)	Ash (%)	Energy content (Kcal/100 g)
Blattodea (cockroaches)	57.3	29.9	5.31	4.53	2.94	–
Coleoptera (beetles, grubs)	40.69	33.4	10.74	13.2	5.07	490.3
Diptera (flies)	49.48	22.75	13.56	6.01	10.31	409.78
Hemiptera (true bugs)	48.33	30.26	12.4	6.08	5.03	478.99
Hymenoptera (ants, bees)	46.47	25.09	5.71	20.25	3.51	484.45
Isoptera (termites)	35.34	32.74	5.06	22.84	5.88	–
Lepidoptera (butterflies, moths)	45.38	27.66	6.6	18.76	4.51	508.89
Odonata (dragonflies, damselflies)	55.23	19.83	11.79	4.63	8.53	431.33
Orthoptera (crickets, grasshoppers, locusts)	61.32	13.41	9.55	12.98	3.85	426.25

Adopted from Lee et al. [27]

Xiaoming and his colleagues reported that the protein content of 100 different species ranged from 13% to 77% based on their dry matter [25]. In Mexico, an analysis of 87 insect species showed a protein content of 15% to 81% with high digestibility [25]. De Castro and his team reviewed the nutritional value of insects that are frequently consumed and found that the protein content varied widely from 1% to 81% [25]. The bioavailability of insect protein is high and is easily digestible, i.e., 76–96%, which is slightly less than that of beef or egg protein. Therefore, insects can be considered a valuable source of protein in our diet [26]. In Central Africa, insects provided about 50% of dietary protein at one point. Compared to plant protein, insect protein has several nutritional advantages, including total protein levels, essential amino acids, and bioavailability. Certain insect species have high lysine, tryptophan, and threonine contents, which are not found in some plants. Edible insects also contain beneficial nutrients like fats with highly unsaturated fatty acids, vitamins, and minerals. Among insects, fat is the second most abundant nutrient (approximately 10–60%, based on dry matter) after proteins. Fats can be categorized into 80% of triglycerides and 20% of phospholipids, which play a role in energy reserves, cell membrane structure, and regulatory physiology. The unsaturated fatty acid profile in edible insects is similar to that of poultry and fish; however, insects have more polyunsaturated fatty acids. Most insects do not contain significant amounts of major omega-3 fatty acids like eicosapentaenoic acid and docosahexaenoic acid, yet their levels could be increased by modifying the feed during insect rearing. Additionally, edible insects are rich in Fe, Zn, Na, Ca, P, Mg, Mn, Cu, riboflavin, pantothenic acid, and biotin. Table 2.1 illustrates the nutritional composition of various edible insects, as reported by Lee et al. 2020.

The advantages of eating edible insects extend beyond their high nutritional content. They have a high feed-to-meat conversion rate and require less land, water, and feed. They also have high fecundity rates with year-round breeding and require little space. In some species (e.g., palm weevil larvae), the byproducts can be used for other livestock and/or humans, which results in a high recycling capability.

2.2.4 Lab-Grown or Cultured Meat Alternatives

Cultured meat, also known as in vitro meat, synthetic meat, lab-grown meat, bioartificial muscle, or Frankenstein meat, represents the latest advancement in the realm of meat alternatives [28]. It is defined as artificial meat produced using stem cell technology. Winston Churchill proposed cultured meat in 1932, the former Prime Minister of the UK. While cell and tissue engineering techniques were originally developed for medical purposes, recent technological advancements have facilitated their application in food technology, enabling large-scale culturing. The first beef patty cultured from bovine muscle cells was introduced to the public in 2013. However, despite early breakthroughs, cultured meat has not been commercialized due to technical challenges in mass production and high costs.

Cultured meat has many advantages, such as its close resemblance to traditional meat produced from farm animals and its potential for environmental sustainability. This product meets the nutritional and sensory preferences of consumers, with superior taste and texture when compared to other meat alternatives. It appeals to consumers who want to maintain their traditional meat consumption habits. During cultured meat production, a single cell can be used to create multiple meat products. This reduces the need for large numbers of animals in livestock farming. Cultured meat can also be utilized for various applications, such as creating functional and designer meat, facilitating quick production, and providing access to exotic and vegan meat options. Additionally, cultured meat offers efficient nutrient and energy conversion, garnering public support and contributing to animal welfare. It can also help to reduce the risk of zoonotic and foodborne diseases, minimize resource use and ecological footprint, and support reforestation and wildlife protection efforts. Furthermore, cultured meat holds the potential for use in space missions and settlements. While the development of cultured meat is ongoing, it holds promise for controlling product ingredients to enhance health benefits without lengthy farming processes. All processes involved in culturing meat adhere to strict sterile conditions and employ various food quality and safety management systems, including Good Manufacturing Practice and Hazard Analysis and Critical Control Points. Thus, cultured meat production can yield safer products devoid of hazards such as contamination, antibiotic abuse, infectious diseases, and food poisoning.

Recent reviews on the production of cultured meat from stem cells have provided valuable insights into various aspects of this emerging technology. Table 2.2 illustrates a summary of recent reviews as presented by Ismail et al. [8].

2.3 The Science Behind Creating Realistic Textures, Flavors, and Nutritional Profiles in Meat Substitutes

Processing and texturization are essential in the development of alternative meats. Plant-based meats usually begin with raw ingredients like soy, peas, or wheat gluten. These ingredients undergo processing techniques such as extrusion, where they

Table 2.2 A summary of recent reviews on producing cultured meat from stem cells and its conclusions

Reference study	Conclusion
Langelaan et al. [39]	The review discusses the requirements to make meat production feasible in vitro. The challenges that need to be addressed include finding an appropriate stem cell source, cultivating them in a 3D environment inside a bioreactor, and providing essential cues for their proliferation and differentiation.
Stephens [40]	The discussion on the development of lab-grown meat and its impact on social, ethical, and regulatory issues.
Bhat and Fayaz [41]	The authors noted that in vitro meat offers health and environmental advantages. However, this technology faces significant challenges in producing highly structured meat and establishing sustainable production on an industrial scale.
Tuomisto and de Mattos [42]	Cultured meat reduces environmental impact by using up to 99% less land, 96% less greenhouse gas emissions, and 82–96% less water than conventional meat.
Dodson et al. [43]	The authors focused on the potential of adipose tissue as a source of stem cells for future clinical applications. Adipose tissue could also contribute to the palatability of cultured meat.
Post [34]	The use of in vitro culturing is being considered as an alternative for producing livestock meat. This bio-artificial muscle has the potential to be a great source of animal protein. However, one of the challenges is to ensure that in vitro meat closely resembles conventional meat in terms of physical properties such as appearance, smell, texture, and taste.
Welin et al. [44]	A review of the ethical issues of cultured meat. The authors concluded "it will need a bit of time to obtain there, it will take people a long time to adapt."
Young et al. [45]	The authors discussed the health benefits of the compounds found in novel meat. They emphasized the availability of essential amino acids, phytanic acid, conjugated linoleic acids, and antioxidants, which are important for a healthy human diet.
Goodwin and Shoulders [46]	A study in some countries about stories of news articles in relation to cultured meat. This discussion can assist scientists in understanding the consumers' perceptions and strategies for the development of cultured meat.
Post [37]	The author presented a proof-of-concept for growing meat in a laboratory, which opens the door for future research on environmentally friendly meat production without the need to raise animals.

Adopted from Ismail et al. [8]

are mixed, cooked, and shaped into meat-like textures using specialized equipment. Texturization is achieved through methods like high-moisture extrusion, which creates fibrous structures similar to meat. Flavoring agents, colors, and binders are often added to enhance taste, appearance, and texture. Fungi-based meats utilize protein-rich sources such as mycelium or fungal biomass. Processing involves extracting proteins from fungi and combining them with other ingredients to form meat analogues. Texturization techniques include extrusion, where the mixture is forced through a die to create meat-like shapes, or fermentation, which produces fibrous structures similar to meat. Insect-based meats involve harvesting edible insects, which are then processed into various forms such as powders, pastes, or whole pieces. Processing techniques may include drying, grinding, or milling the insects into fine particles, which are then mixed with other ingredients to form meat-like products. Texturization can be achieved through methods such as extrusion or molding, resulting in products with diverse textures and flavors. Lab-grown meats are produced by harvesting stem cells from animals and culturing them in a laboratory setting. These cells are then grown in a nutrient-rich medium and stimulated to differentiate into muscle tissue. Texturization is achieved through tissue engineering techniques, where the cells are scaffolded and layered to mimic the structure of traditional meat. Bioreactors are used to provide the necessary environment for cell growth and maturation.

2.3.1 Texturization, Appearance, and Binding in Plant-Based Meat Alternatives

Consumer satisfaction hinges on the consistency, taste, and appearance of meat alternatives. By carefully selecting and blending ingredients, manufacturers can create products that meet consumer expectations while offering sustainable and ethical alternatives to traditional meat. Meat alternatives such as plant-based burgers or sausages rely on the selection and formulation of ingredients to provide the desired sensory and textural characteristics. These products typically contain around 50–80% water, 4–20% non-textured protein, 10–25% vegetable textured proteins, 3–10% flavor enhancers, 0–15% fats, 0–5% coloring agents, and 1–15% binding agents [29]. The high water content helps to reduce costs, impart juiciness, soften the product during processing, and aid in emulsification. Protein additions contribute to texture, taste, and physical appearance, with textured proteins playing a key role in mimicking meat-like textures. These proteins can either be mixed with meat or completely replaced to create vegetarian and vegan products. While meat extenders alone may not replicate the texture, appearance, or taste of meat, they enhance overall product quality when combined with meat. Meat alternatives are specifically formulated to mimic the form, texture, taste, and color of whole meat without using any meat-derived components. Achieving the desired consistency often requires minimal use of chemicals or additives. To improve the water holding capacity, ingredients such as soy protein concentrates, egg whites, wheat gluten, or adhesive agents like starches and hydrocolloids are commonly employed.

The demand for plant-based proteins is on the rise due to cost, accessibility, and physiological properties. Proteins have important functional properties like water and oil retention, emulsifying, gelation, and foaming. The quality and functional properties of proteins help to build meat-like structure and consistency. Soy protein is mostly used due to its desired traits and low cost, but proteins from other sources like cereals, oil-seed plants, fungi, micro-algae, and fermentation are also used for making meat substitutes. Soy proteins are commonly used in meat alternatives due to their functional properties. They can be added as flour, isolates, or concentrates. Soy flour is the least processed, while isolates have the highest protein content at 90%. Soy protein isolate has a bland flavor and light color, making it desirable for meat analogues. However, additional components can be beneficial for meat replacements. The gluten from wheat forms fibrous structures due to protein desulfation bonding. Hydrogen bonds, disulfide bonds, and hydrophobic interactions are responsible for the formation of meat-like structures. Wheat gluten produces desirable dough for the development of meat analogues. Legume proteins like lentils, chickpeas, lupine, peas, and different beans have been studied for creating gels, stabilizing foams, and emulsifying properties. Pea protein is the most viable option for meat substitutes, and pre-treating legumes can influence gelatinization activity. Protein isolates with high protein content increase hardness and chewiness in meat analogues. Fibrous meat analogues can be developed by combining oat fiber and pea protein isolate. Peanut is a major oilseed with a global production of 50.61 million tons. Its defatted protein content is around 50% and can be obtained as a by-product from the oil extraction industry. Peanut protein has an excellent amino acid profile, no flatulence factor, and bean flavor. It can be used as a ground beef substitute in snack foods and other meat-based food formulations. However, the poor gelling ability of peanut protein compared to soy protein can be improved by adding polysaccharides and transglutaminase. Peanut allergy is still a major concern, but combining transglutaminase modification and high-moisture extrusion can significantly reduce peanut allergens in these products. Various proteins from sources such as rice, pseudocereals, beans, and vegetables have been used as alternative protein sources in combination with soy or pea proteins for plant-based meat substitutes. Rice protein has poor water solubility and a bitter flavor but can be texturized by combining it with soy protein. Hemp protein can substitute up to 60% of soy in high-moisture extruded meat substitutes. Oat protein has a superior amino acid profile, and its fibrous structure formation can be improved by enzymatic modification. Mung bean protein has a lower carbon footprint and an amino acid profile resembling soybean protein. Lupin protein can be used to replace egg or improve protein content but has weaker gel strength than soy or pea proteins. Faba bean protein is a good source of lysine and can be texturized through high-moisture extrusion. Table 2.3 illustrates common proteins used in the commercial production of plant-based alternative meats as described by Wang et al. [7].

Canola and rapeseed are recognized as valuable protein sources with notable emulsifying and foam formation properties, as documented by Chang et al. [30] and Ntone et al. [31]. Under conditions of high pressure or heat, rapeseed proteins exhibit gelatinization behavior, leading to the development of a texture akin to that of meat. Moreover, napin, a component found in rapeseed, has emerged as a potential extender for casein-containing compounds. Napin has the ability to form aggregates with β-casein through various bonding mechanisms, including salt-bridging, ionic interactions, and hydrophobic bonding, as elucidated by Schwartz et al. [32]. On the other hand, canola proteins are predominantly composed of globulins or

Table 2.3 Common proteins used in commercial plant-based meat alternatives

Proteins	Composition	Functionalities	References
Soybean protein	Isolate and concentrate; storage globulins mainly glycinin (11S, hexamer, MW 320–380 kDa) and β-conglycinin (7S, trimer, MW 150–220 kDa); basic and acidic subunits in glycinin linked by disulfide bonds.	Aggregation, gelation, and fiber formation through heating and extrusion; oil-binding and emulsification	Fukushima (2011)
Pea protein	Isolate and concentrate; predominant globulins: Legumin (11S, hexamer, MW 320–380 kDa), vicilin (7S, trimer, MW 150–170 kDa, lacks cysteine residues; minor fraction conviction (MW 290 kDa).	Functionality like soy proteins	Barac et al. (2010), Lam et al. (2018)
Wheat protein	Gluten is an elongated structure with two components: gliadin (alcohol-soluble, polypeptides with MWs 25–100 kDa linked by intramolecular disulfide bonds) and glutenin (alkali-soluble, polymeric large subunits MW66–88 kDa and small subunits MW 32–45 kDa linked into polymers of 150–1500 kDa)	Used together with legume protein to provide fibrous texture; elasticity and extensibility	Shimoni and Galili (1996)

albumin fractions, which have been shown to form cohesive gel-like structures under certain salt concentrations. In combination with k-carrageenan, canola protein exhibits robust and flexible connections, suggesting its suitability as a structural building agent. These findings highlight the versatility of canola and rapeseed proteins in contributing to the textural properties and structural integrity of food products, particularly in the context of meat alternatives. Meat replacements often have low fat as defatted materials are commonly used. Adding oil or fat during processing can impact the fibrous structure formation, which affects macromolecule alignment. Adding vegetable oil and fat can enhance taste, softness, and flavor. Moderately fermented soy proteins can be added to meat substitutes as a replacement or supplement. Canola oil, rapeseed oil, soya oil, palm oil, coconut oil, and sunflower oil are some commonly used fats for plant-based meat replacements. Adding oil or fat is vital as it increases the taste of meat alternatives.

Various ingredients are added during the manufacturing process of plant-based meat analogues to enhance their texture and flavor. Table 2.4 outlines the specific techniques for improving these qualities [18]. Since plants mainly consist of amorphous tissue, several methods have been employed to address texture concerns, such as spinning, thermoplastic extrusion, and steam texturization. Among these, extrusion is the most widely used technique due to its cost-effectiveness and ability to produce meat analogues in various shapes and sizes. This process involves a screw system within a barrel, where plant proteins undergo compression and heating to restructure into a striated, layered, and cross-linked mass, ultimately producing textured vegetable protein (TVP) [33]. Previous studies have shown that incorporating wheat gluten and soybean protein in TVP formulations can give a similar

Table 2.4 Ingredients used in the production of plant-based meat substitutes

Ingredient	Purpose	Usage level (%)
Water	Ingredient distribution	50–80
	Emulsification, juiciness, cost	
Textured vegetable proteins	Water binding, texture/mouthfeel	10–25
	Appearance; protein fortification/nutrition	
	Source of insoluble fiber	
Non-textured proteins	Water binding, emulsification	4–20
	Texture/mouthfeel	
	Protein fortification/nutrition	
Flavors/spices	Flavor: Savory, meaty, roasted, fatty, serumy	3–10
	Flavor enhancement (for example, salt)	
	Mask cereal notes	
Fat/oil	Flavor, texture/mouthfeel	0–15
	Succulence, Maillard reaction/browning	
Binding agents	Texture/"bite," water binding, may contribute to fiber content, can determine production processing conditions	1–5
Coloring agents	Appearance/eye appeal	0–0.5
	Natural or artificial	

appearance, texture, taste, and nutritional value to conventional meat. Another protein source is mycoprotein, which is derived from starch by-products using fungi and has a structure and diameter similar to muscle fibers found in meat. This results in a texture that closely resembles traditional meat products. These techniques help develop plant-based meat analogues that meet consumer expectations for texture and flavor, enhancing their acceptance and marketability.

Binder components in meat substitutes can be derived from plants or animals and act as both moisture and fat binders. Wheat gluten, xanthan gum, soy protein isolate, eggs, and carrageenan are commonly used binding agents. The quality and nutritive values of plant-based substitutes depend on the type and concentration of binding agents. To mimic the flavor of meat, savory aromas, spices, and precursors are added. Thermal treatments like extrusion can affect the flavor quality, making the optimization of taste and flavor a complex process. Several compounds, including amino acids, nucleotides, and thiamine, are used to duplicate meat aromas in meat substitutes. Color is crucial in meat and meat alternatives. Additives such as caramel colors, annatto, and betadine are used to achieve the desired appearance. Heat-labile colorants and reducing chemicals are added to create a raw and cooked appearance. Adjusting the pH using citric acid, acetic acid, or lactic acid can help with color accuracy. Color retention aids such as maltodextrin and hydrated alginate can be used to control color migration.

2.3.2 *Mycoprotein-Based Meat Alternatives—Texturization and Presentation*

Mycoprotein, a valuable source of single-cell protein (SCP), is predominantly produced through fermentation processes involving special microorganisms such as fungi, yeasts, bacteria, and algae cultivated in suitable media [34]. Among these microorganisms, *Fusarium venenatum* stands out as the primary source of mycoproteins, possessing high protein biomass content and receiving market approval. In the fermentation process for SCP production, factors such as environmental sustainability, cost-effectiveness, and safety considerations are paramount. Fermentation media typically consist of food-grade carbohydrates, potassium, magnesium, phosphorus, ammonium salts, trace elements, and vitamins like biotin, essential for mycelial growth. Various fermentation systems, including solid-state, semisolid, submerged, and surface cultures, can be employed, with submerged fermentation yielding higher yields. After fermentation, the broth undergoes treatment, typically heating, to reduce RNA content. Biomass separation techniques such as centrifugation or filtering yield pastes resembling chicken breast texture or bread dough. Fungi like *F. venenatum* can thrive on synthetic media containing ammonium, glucose, and biotin, though the total cost of mycoproteins depends on the substrates used. Agro-industrial byproducts offer a cost-effective alternative, with pea-processing byproducts showing promise due to their high protein content compared to synthetic media. Various substrates, including fruit juices, carbohydrates, waste materials, date wastes, molasses, starches, fruit and vegetable wastes, and natural gases, can be utilized for mycoprotein production. Molasses, for instance, are readily available, cost-effective, and free from toxic substances and fermentation inhibitors. Cassavas, rich in starches, are also suitable substrates. Yields of protein production may vary depending on the substrates and microbial species employed. Ultimately, the availability of substrates and proximity to production plants significantly influence the success of mycoprotein production processes.

Quorn minces and pieces are produced through a continuous batch method. To make both forms, mycoproteins are treated with small amounts of egg albumen and water. Malt extract is added to make Quorn mince, while Quorn pieces receive a natural flavoring to enhance their taste. Mixing is done at low shear in batches of two tons. The process can be challenging due to the complex hydration properties of egg albumen and the viscoelastic nature of mycoprotein, but it is simplified to mitigate difficulties. Additionally, a small amount of calcium is included to enhance interhyphal crosslinking and achieve a meat-like texture in the final products. The mixed mass is then transferred to forming equipment to shape blocks under pressure. Afterwards, live steam is used to raise the temperature to 85–90 °C to denature the proteins. The products are cooled to −18 °C for 30 min, which is a crucial step in creating a meaty texture. Controlled growth of ice crystals during freezing efficiently binds filamentous hyphae, forming fibrous bundles with a meat-like texture. The stability of both the production process and final products relies on the strength of mycoproteins and egg networks, along with the gelled proteins and entangled

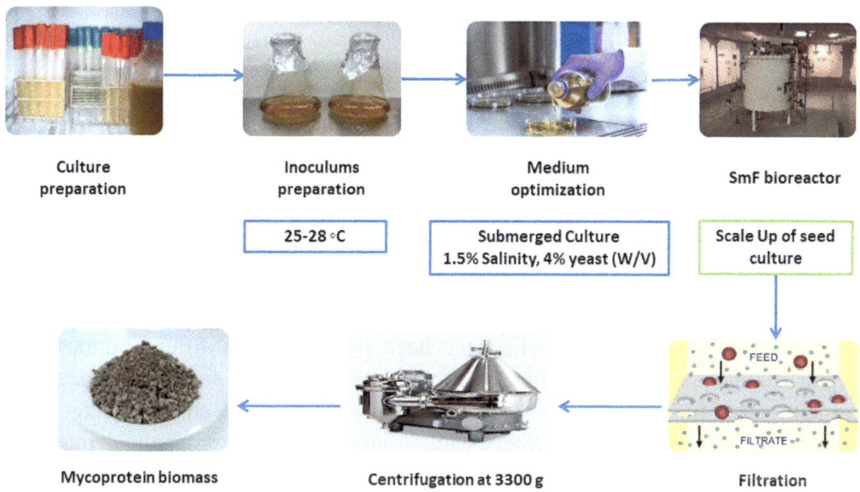

Fig. 2.2 Flow chart of crucial steps in mycoprotein production as meat substitutes [34]

fibers. Figure 2.2, adopted from Baltork et al. (2020), illustrates the major steps involved in mycoprotein production and the resulting products.

2.3.3 Insect Meat—Appearance and Presentation

Incorporating insects into the human diet is a promising approach, and using insect-based ingredients and products can help promote their acceptance [24]. By converting insects into familiar forms, such as powders or extracts, food industry can maximize their potential as food resources. To do this, insects go through raw material processing methods like drying and milling to be turned into dry powders. This not only reduces their transportation volume but also increases their shelf life by lowering water activity. Extraction methods for protein and oil have also been explored to enhance the nutritional content and functional properties of insect-derived ingredients, improving the quality of various food products. By adding insect-based ingredients to common foods like bread, cookies, and sausages, we can increase their nutritional value and improve overall food quality. However, safety concerns about edible insects continue to persist. Anti-nutrients such as chitin and potential toxic substances pose challenges that require further investigation to ensure the safety of consuming insect-based foods. Additionally, there are microbial risks and allergens to consider. There is limited research on insect allergies, and comprehensive studies are necessary to address potential allergic reactions and cross-reactive proteins, especially in arthropods like arachnids and crustaceans. Addressing these safety considerations is critical for the sustainable growth and acceptance of the edible insect industry.

2.3.4 Sensorial Aspects of Cultured Meat

The process of creating cultured meat involves taking a small piece of tissue from a living animal via a biopsy conducted under anesthesia. Skeletal muscle cells are cultured from satellite cells, which go through the proliferation and differentiation phases. During the proliferation stage, stem cells are isolated and encouraged to develop into new muscle tissue. These stem cells can renew themselves and generate new tissue as long as growth factors are present. The cultured cells are kept in a specific liquid media that contains tailored nutrients for tissue growth. The proliferation process doubles the cell population every 7–8 weeks until trillions of cells are produced. Once enough cells have been produced, the differentiation stage begins. Satellite cells differentiate in the absence of growth factors and naturally come together to form myotubes. These cells are then embedded in a collagen gel, forming donut-shaped muscle fibers with a diameter of approximately 1 mm. The innate contractile ability of muscle cells serves as a potent stimulus for muscle maturation and protein production. The production of these muscle fibers is relatively rapid, taking only 3 weeks to harvest. Approximately 10,000 muscle fibers are required to prepare 85 g of hamburger. Figure 2.3 summarizes the process flow in producing cultured meat.

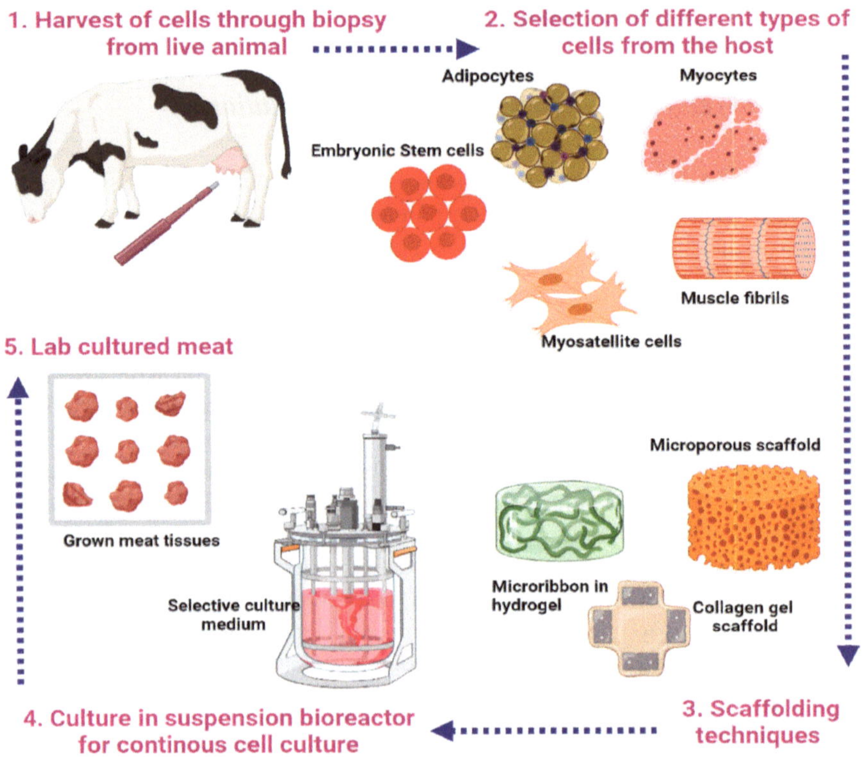

Fig. 2.3 Process flow-developing cultured meat [35]

Cultured meat production encounters different challenges depending on the type of meat being produced. Ground or finely minced meat products are easier to produce, unlike mimicking the texture of full-sized meat cuts, which presents significant hurdles. Achieving a steak-like texture requires a perfusion system and the assembly of a vascular-like system. Emulating traditional meat's texture requires co-culturing with different cell types or using an edible scaffold made of structural proteins. The use of cultured meat in ground meat products is more feasible, but additional ingredients such as breadcrumbs, egg white powder, and binders may be necessary to achieve the desired texture. Synthetic meat proteins produced through fermentation could offer a more feasible alternative to entire muscle cells, especially for finely minced meat products. Cultured meat usually has a pale color due to low myoglobin expression under ambient oxygen conditions. Methods like adding lipids or acetic acid to the culture medium, supplementing it with extra iron, or culturing under low oxygen conditions can increase myoglobin content. Fresh meat's taste is influenced by compounds like lactate and IMP, Maillard reaction, and lipid degradation upon cooking. Fat contributes significantly to meat flavor, texture, and juiciness. Co-culturing with adipocytes or using plant-based fats can add fat to cultured meat. Artificial flavors may be added if necessary. Technological innovation and ingredient optimization are necessary to achieve consumer satisfaction.

References

1. Ketelings L, Havermans RC, Kremers SPJ, de Boer A. How different dimensions shape the definition of meat alternative products: a scoping review of evidence between 2000 and 2021. Curr Dev Nutr. 2023;7(7):101960. https://doi.org/10.1016/j.cdnut.2023.101960.
2. Ferreira JP, Sharma A, Zannad F. The future of meat: health impact assessment with randomized evidence. Am J Med. 2021;134(5):569–75. https://doi.org/10.1016/j.amjmed.2020.11.007.
3. Ruiz-Capillas C, Herrero AM. Sensory analysis and consumer research in new product development. Foods. 2021;10(3):582. https://doi.org/10.3390/foods10030582.
4. Kurek MA, Onopiuk A, Pogorzelska-Nowicka E, Szpicer A, Zalewska M, Półtorak A. Novel protein sources for applications in meat-alternative products-insight and challenges. Foods. 2022;11(7):957. https://doi.org/10.3390/foods11070957.
5. Lima M, Costa R, Rodrigues I, Lameiras J, Botelho G. A narrative review of alternative protein sources: highlights on meat, fish, egg and dairy analogues. Foods. 2022;11(14):2053. https://doi.org/10.3390/foods11142053.
6. Stephens N, Di Silvio L, Dunsford I, Ellis M, Glencross A, Sexton A. Bringing cultured meat to market: technical, socio-political, and regulatory challenges in cellular agriculture. Trends Food Sci Technol. 2018;78:155–66. https://doi.org/10.1016/j.tifs.2018.04.010.
7. Wang Y, Jian C. Novel plant-based meat alternatives: implications and opportunities for consumer nutrition and health. Adv Food Nutr Res. 2023;106:241–74. https://doi.org/10.1016/bs.afnr.2023.03.006. Epub 2023 Jun 14
8. Ismail I, Hwang YH, Joo ST. Meat analog as future food: a review. J Anim Sci Technol. 2020;62(2):111–20. https://doi.org/10.5187/jast.2020.62.2.111.
9. Szenderák J, Fróna D, Rákos M. Consumer acceptance of plant-based meat substitutes: a narrative review. Foods. 2022;11(9):1274. https://doi.org/10.3390/foods11091274.

10. Bryant CJ. Plant-based animal product alternatives are healthier and more environmentally sustainable than animal products. Future Foods. 2022:100174. https://doi.org/10.1016/j.fufo.2022.100174.
11. Font-I-Furnols M. Meat consumption, sustainability and alternatives: an overview of motives and barriers. Foods. 2023;12(11):2144. https://doi.org/10.3390/foods12112144.
12. Smetana S, Ristic D, Pleissner D, Tuomisto HL, Parniakov O, Heinz V. Meat substitutes: resource demands and environmental footprints. Resour Conserv Recycl. 2023;190:106831. https://doi.org/10.1016/j.resconrec.2022.106831.
13. Petersen T, Hirsch S. Comparing meat and meat alternatives: an analysis of nutrient quality in five European countries. Public Health Nutr. 2023;26(12):3349–58. https://doi.org/10.1017/S1368980023001945.
14. Reid Y, Denis F, Hatia X, Tizhe F. Right to Food Assessment of the Main EU Food Laws and Policies. 2021. Available at SSRN 4058466.
15. Rubio NR, Xiang N, Kaplan DL. Plant-based and cell-based approaches to meat production. Nat Commun. 2020;11(1):6276. https://doi.org/10.1038/s41467-020-20061-y.
16. Kenny TA, Woodside JV, Perry IJ, Harrington JM. Consumer attitudes and behaviors toward more sustainable diets: a scoping review. Nutr Rev. 2023;81(12):1665–79. https://doi.org/10.1093/nutrit/nuad033.
17. Meat Substitutes Market Size, Growth, Potential, Report, 2031. Retrieved March 8, 2024, from https://www.fortunebusinessinsights.com/industry-reports/meat-substitutes-market-100239.
18. Ahmad M, Qureshi S, Akbar MH, Siddiqui SA, Gani A, Mushtaq M. Plant-based meat alternatives: compositional analysis, current development and challenges. Appl Food Res. 2022;2(2):100154. https://doi.org/10.1016/j.afres.2022.100154.
19. Benny A, Pandi K, Upadhyay R. Techniques, challenges and future prospects for cell-based meat. Food Sci Biotechnol. 2022;31(10):1225–42. https://doi.org/10.1007/s10068-022-01136-6.
20. Arora S, Kataria P, Nautiyal M, Tuteja I, Sharma V, Ahmad F. Comprehensive review on the role of plant protein as a possible meat analogue: framing the future of meat. ACS Omega. 2023;8(26):23305–19. https://doi.org/10.1021/acsomega.3c01373.
21. Sethi S, Tyagi SK, Anurag RK. Plant-based milk alternatives an emerging segment of functional beverages: a review. J Food Sci Technol. 2016;53(9):3408–23. https://doi.org/10.1007/s13197-016-2328-3. Epub 2016 Sep 2
22. Souza Filho PF, Andersson D, Ferreira JA, Taherzadeh MJ. Mycoprotein: environmental impact and health aspects. World J Microbiol Biotechnol. 2019;35(10):147. https://doi.org/10.1007/s11274-019-2723-9.
23. Collins CM, Vaskou P, Kountouris Y. Insect food products in the Western world: assessing the potential of a new 'green' market. Ann Entomol Soc Am. 2019;112(6):518–28. https://doi.org/10.1093/aesa/saz015. Epub 2019 Sep 11
24. Ros-Baró M, Casas-Agustench P, Díaz-Rizzolo DA, Batlle-Bayer L, Adrià-Acosta F, Aguilar-Martínez A. Edible insect consumption for human and planetary health: a systematic review. Int J Environ Res Public Health. 2022;19(18):11653. https://doi.org/10.3390/ijerph191811653.
25. Rodríguez-Rodríguez M, Barroso FG, Fabrikov D, Sánchez-Muros MJ. In vitro crude protein digestibility of insects: a review. Insects. 2022;13(8):682. https://doi.org/10.3390/insects13080682.
26. Skotnicka M, Karwowska K, Kłobukowski F, Borkowska A, Pieszko M. Possibilities of the development of edible insect-based foods in Europe. Foods. 2021;10(4):766. https://doi.org/10.3390/foods10040766.
27. Lee JH, Kim T, Kim YJ, Kang M, Lee HG, Kim BK, Choi Y. Structural, physicochemical, and immune-enhancing properties of edible insect protein isolates from Protaetia brevitarsis larvae. Food Chem X. 2023;18:100722. https://doi.org/10.1016/j.fochx.2023.100722.
28. Anang DM, Fuseini A. Production of cultured meat: challenges and opportunities. In: Future proteins: sources, processing, applications and the bioeconomy. p. 285–310. https://doi.org/10.1016/B978-0-323-91739-1.00014-3.

29. Kyriakopoulou K, Keppler JK, van der Goot AJ. Functionality of ingredients and additives in plant-based meat analogues. Foods. 2021;10(3):600. https://doi.org/10.3390/foods10030600.
30. Chang C, Tu S, Ghosh S, Nickerson MT. Effect of pH on the inter-relationships between the physicochemical, interfacial and emulsifying properties for pea, soy, lentil and canola protein isolates. Food Res Int. 2015;77:360–7. https://doi.org/10.1016/j.foodres.2015.08.012.
31. Ntone E, Qu Q, Gani KP, Meinders MBJ, Sagis LMC, Bitter JH, Nikiforidis CV. Sinapic acid impacts the emulsifying properties of rapeseed proteins at acidic pH. Food Hydro. 2022;125:107423. https://doi.org/10.1016/j.foodhyd.2021.107423.
32. Schwartz JM, Solé V, Guéguen J, Ropers M, Riaublanc A, Anton M. Partial replacement of β-casein by napin, a rapeseed protein, as ingredient for processed foods: Thermoreversible aggregation. LWT Food Sci Tech. 2015;63(1):562–8. https://doi.org/10.1016/j.lwt.2015.03.084.
33. Malav OP, Talukder S, Gokulakrishnan P, Chand S. Meat Analog: a review. Crit Rev Food Sci Nutr. 2013;55(9):1241–5. https://doi.org/10.1080/10408398.2012.689381.
34. Majumder R, Miatur S, Saha A, Hossain S. Mycoprotein; production and nutritional aspects: a review. Sustainable Food Technology. 2024;2(1):81–91. https://doi.org/10.1039/D3FB00169E.
35. Soleymani S, Naghib SM, Mozafari MR. An overview of cultured meat and stem cell bio-printing: how to make it, challenges and prospects, environmental effects, society's culture and the influence of religions. J Agric Food Res. 2024;22:101307. https://doi.org/10.1016/j.jafr.2024.101307.

Chapter 3
Environmental Impact of Meat Substitutes

The domain of food production is one of the most environmentally impactful human activities. Farming activities have a significant environmental impact, accounting for 61–81% of greenhouse gas emissions (GHGE), 79% of acidification, and 95% of eutrophication related to food impacts [1]. The challenge of nourishing a projected population of 10 billion by 2050 and increasing the calorific energy supply from 30 to 45 exajoules poses a formidable obstacle to conventional high-calorie diets prevalent in high-income countries. These diets predominantly rely on heavily processed and animal-derived products [2]. Adhering to the current approach for feeding the global population would result in the depletion of natural resources and almost double the rates of overweight and obesity [3].

Meat production, often accused of substantial environmental impacts, exemplifies this concern. Livestock alone contributes one-third of the current human-induced nitrogen emissions or 65 Tg N year^{-1}. [4]. If current trends in meat consumption persist, predictions indicate that by 2030, it will be responsible for 37% and 49% of the greenhouse gas budget permitted under the 2 °C and 1.5 °C targets, respectively [5]. Envisioning adherence to sustainability targets, it becomes imperative to shift diets towards reducing meat consumption by more than 50% and increasing plant food consumption by more than 100% [6]. Despite animal-derived products supplying only 17% of the world's food and approximately 40–58% of its proteins, the environmental effects of animal production are disproportionately high. Animal agriculture takes up 77% of all agricultural lands, utilizes 30% of all water resources, and contributes 12–20% of human-induced GHGE [7]. Moreover, the environmental impact extends to eutrophication, particularly pronounced on a regional scale due to the role of animal manure [8].

Despite their considerable environmental impact, animal-derived products assume a pivotal role in both the economic and cultural spheres of society [9]. Market analysts estimate the global value of this sector to exceed 1 trillion USD, with the US market constituting a substantial portion at 838 million USD. The meat

S. Ray, S. Suri, *The Sustainable Plate: Unveiling the Science of Meat Substitutes and Impact on Global Health*, https://doi.org/10.1007/978-3-031-94604-2_3

industry is projected to experience a 20% growth by 2025. Beyond economic and traditional considerations, meat serves as a valuable source of nutrients, playing a crucial role in current diets by contributing 28 grams of protein and 30% of daily caloric intake in European diets [10]. However, the rapid surge in meat consumption over the past two decades poses a risk of adverse consequences for future generations in terms of resource depletion and environmental degradation. While vegetarian and vegan diets, recognized for their lower environmental impact, are gaining popularity, only around 5% of the global population identifies as such, with the majority adopting a flexitarian approach—occasionally consuming meat [11].

Meat products are vital contributors to society, providing essential proteins, amino acids, and microelements, and they significantly influence the sustainability of diets, prompting discussions on the imperative for low environmental impact diets in the future [12]. The predominant issue associated with meat consumption is not the act itself but rather the alarming rates of overconsumption. The correlation between elevated meat consumption, particularly processed red meat, and heightened risks of cancer and metabolic diseases is well-established. Notably, the type of meat, processing methods, and individual lifestyle play critical roles in determining the associated health risks. For instance, synergistic and antagonistic effects between the consumption of meat products and nutrients exist. Consuming foods high in fiber, calcium, selenium, and Vitamins C, D, and E can potentially offset the negative cancer-causing effects of meat production [13]. Recommendations to limit red meat consumption to 25–70 grams per day have been proposed to mitigate associated health risks [14]. Current consumption patterns in Western countries depict a fivefold increase over the past 20 years ago, with rates that are eight times greater than those in developing countries. This surge in overconsumption and overproduction not only leads to health issues such as obesity, high blood pressure, and heightened carcinogenic risks but also contributes significantly to environmental impacts [15]. To transition the food system toward a healthier state, considering both nutritional and environmental aspects, consumers are increasingly seeking for meat substitutes that let them enjoy the meat's flavor without having negative impacts on their health or the environment.

The concept of meat substitution remains somewhat nebulous, potentially rooted in the historical evolution of the requirement for protein sources and later to replace meat. Clarifying terminology used for meat substitutes is crucial. "Meat alternative" serves as a broad term encompassing any protein source—plant, animal, fungi, or microalgae—that can replace meat in a meal [16]. This term, closely associated with "alternative protein," primarily emphasizes the necessity of protein supply without mandating precise mimicry of all nutritional and textural properties [17]. Conversely, the terms "meat analogue" or "meat substitute" are a more specific descriptors for products that replicate meat characteristic related to processing, nutritional content, and sensory attributes [18]. Although meat analogues are frequently associated solely to plant biomass and texturized vegetable protein (TVP) technologies, this classification overlooks several meat analogues on the market, including those based on insects and microalgae [19]. "Meat analogue" encompasses a diverse range of products, further differentiated based on intended application (processing functionality) into: (1) meat analogues that imitate whole muscle tissue, (2) meat preparation analogues that imitate fragmented whole muscle tissue (e.g., minced meat), and (3) processed meat analogues that imitate processed meat products (e.g., sausage) [20]. This review employs the terms "meat analogues" and "meat substitute" interchangeably and describes imitations of meat that are physically, enzymatically, or biologically structured, made from proteins, fats, carbohydrates, and other substances derived from non-animal sources and less frequently used animal species, irrespective of the level of processing.

Historically, there were few primary criteria that were followed when substituting different protein sources for meat, the first of which was related to local or regional availability. This consideration was paramount even before the era of globalization. Local biomass's high protein content led to the development of substitutes such as tofu, tempeh, fermented breadfruit products, jackfruit, oncom, seitan, mushrooms, and other protein cheeses, as affordable and accessible alternatives. However, globalization subsequently increased meat availability in many regions worldwide, prompting concerns about the necessity for healthy diet and reducing meat consumption. The availability of meat led to the development of new alternatives, emphasizing the ability to replicate texture and the taste of meat, coupled with requirements for improved sustainability. These considerations led to the development of processing technologies aiming to mimic meat texture, as well as the replication of meat itself under controlled conditions [21]. While criteria associated with abundance, economic feasibility, social considerations, and techno-functional soundness are well-explored in the scientific literature [22], evaluation of the sustainable effects, particularly environmental ones, of meat substitutes from a holistic perspective remains infrequent. Therefore, the goal of this review is to organize the latest available knowledge on the resource requirements and environmental effects of meat substitutes and analogues.

It is imperative to reduce global greenhouse gas emissions [23] and adhere to the two degrees Celsius limit stipulated in the Paris Climate Agreement [19], aiming to alleviate the ongoing climate crisis and forestall more severe consequences. As highlighted by Poore and Nemecek [24], food contributes a quarter to the global carbon footprint. Given that meat production is a substantial contributor to problematic greenhouse gases in the food sector [24, 25], opting for meat alternatives emerges as a viable choice to substitute traditional meat, thereby aiding in climate conservation [26]. The substantial land use linked to animal husbandry and feed production poses environmental challenges. Nearly half of the world's habitable land is designated for agriculture, with 77% utilized as grazing land for animal feed or feed production [27]. This implies that only a fraction of the agriculturally suitable land is currently employed directly for cultivating crops intended for human consumption. From a societal standpoint, the continual surge in global meat consumption raises ethical and health-related apprehensions. Since 1960, worldwide meat production has surged fivefold [28]. Despite the well-documented health risks associated with high meat consumption, such as an increased incidence of type two diabetes, colorectal cancer, coronary heart disease, and various cardiovascular and zoonotic diseases [29–31], global meat consumption remains on an upward trajectory.

Among the various factors contributing to the increasing demand for meat alternatives, one potential reason could be the growing awareness of the adverse effects associated with diets rich in red and, particularly, processed meats on human health [32]. This awareness, coupled with a heightened concern for the environmental consequences of animal products compared to their plant-based counterparts, encourages the shift towards sustainable and healthy diets, emphasizing a higher intake of plant-based foods and moderate consumption of animal products [33].

To thoroughly explore the potential impact of Plant-Based Meat Alternatives (PBMAs) on human health, it is crucial to scrutinize the nutritional characteristics of these products. This analysis is essential considering that meat serves as a vital source of high-quality proteins, iron, vitamins, minerals, and varying amounts of saturated fats based on the meat type [34]. Several studies have investigated the nutritional quality of meat alternatives in different markets, comparing them with animal-based meat in terms of energy and nutrient content [35–37].

For instance, a recent study assessed the nutritional quality of 269 commercial meat analogues available in the Italian market by examining data on their food labels [38]. The study revealed considerable nutritional variability among PBMAs, with plant-based steaks exhibiting notably higher protein and lower energy, fat, and salt contents compared to other plant-based food categories. When compared to reference animal meat products, all PBMAs showed higher fiber content. Additionally, plant-based burgers and meatballs exhibited lower protein content than their meat counterparts, while ready-sliced meat substitutes demonstrated lower salt content than cured meats. Similar findings were reported in studies conducted in the USA [39], Sweden [40], and other European markets [41]. These studies indicated lower energy, total, and saturated fat contents and higher total carbohydrates, sugars, and fibers in PBMAs compared to meat-based products. However, salt content showed mixed results. Moreover, both plant-based and meat-based products generally presented comparable total protein amounts, despite significant differences in single amino acid contents. Specifically, PBMAs displayed higher levels of glutamic acid and cysteine and lower levels of alanine, glycine, and notably methionine [42].

These results emphasize the need for further exploration of plant-based protein blends to minimize distinctions between plant-based and animal-based meats [43]. Additionally, it is essential to note that plant-based and animal products differ in protein digestibility and the bioavailability of single amino acids. Animal meat demonstrated higher protein digestibility than PBMAs, impacting amino acid bioavailability negatively. This underscores the importance of using specific protein sources with high bioavailability, such as soy isolate, and considering the actual bioavailability of amino acids when assessing the dietary quality of patterns incorporating these products. Another noteworthy consideration concerns micronutrients. While data on this topic are often limited, previous studies have suggested that PBMAs serve as a good mineral source, reporting higher iron content compared to meat [44, 45]. However, it is crucial to highlight that iron absorption and bioavailability from plant-based sources and vegetarian diets are lower than omnivorous diets, a factor that should be addressed in future investigations [46]. In summary, these findings underscore the significance of a careful evaluation of the nutritional implications associated with the transition from animal meat to PBMAs to identify potential nutrient risks [47].

Meat, acknowledged for its high biological value in protein, faces sustainability challenges in its production due to the extensive inputs and utilization of finite natural resources [48]. Presently, diverse meat production systems exist, with production efficiency hinging on factors such as feeding practices, breeds, management, and technology application [49, 50]. In contrast, crop growth demands fewer resources per unit, presenting an enticing avenue for sustainable development amidst escalating food demands in developed nations not reliant on subsistence animal breeding [51]. PBMAs in such contexts could offer environmental advantages concerning biodiversity, land and water utilization, and diminished greenhouse gas (GHG) emissions [52, 53]. However, the environmental ramifications of PBMAs necessitate assessment, often conducted through the life cycle assessment (LCA) approach. This methodology, guided by ISO standards, quantifies a product's environmental impacts and seeks to enhance its overall environmental performance [54, 55]. Multiple LCA studies on PBMAs have been undertaken, examining hotspots in production processes and comparing environmental performances with animal-based products. Indicators like climate change, land use, water consumption, and energy use are pivotal considerations.

For instance, Bryant's analysis of 43 studies concluded that meat analogue production tends to be more sustainable compared to animal products [56]. Detzel et al. proposed that PBMAs could alleviate environmental impacts related to food consumption by optimizing ingredient processing and stabilizing yields of protein sources like legumes [57]. Smetana et al. highlighted the potential of technology, including machinery and process equipment, to enhance the sustainability of alternative protein source production [58]. Mejia et al.'s detailed LCA study on three factories producing 57 meat analogue types demonstrated low GHG emissions primarily attributed to the manufacturing process, agricultural production of ingredients, and transportation [59]. Goldstein et al. emphasized the pivotal role of the production stage, accounting for 80% of the environmental impact due to fossil-based electricity use, suggesting alternative energy solutions could mitigate this impact [60].

Despite these insights, further in-depth studies are imperative, given the contrasting data attributed to energy consumption derived from proxy processes for implemented energy sources [56]. In the meat supply chain, meat production and animal husbandry emerge as the most impactful stages [61]. Manure production, when applied to soil, negates the need for chemical fertilizer, contributing to crop yield and soil fertility. On the contrary, legumes, through nitrogen fixation, eliminate the need for nitrogen fertilization, resulting in lower emissions of N2O and NH3 due to the non-use of manure or synthetic fertilizers [62].

Studies examining the impact of meat and meat analogues on water usage and their effects on eutrophication and acidification reveal varying results. Smetana et al.'s comparison of patties with and without meat indicated lower acidification and subsequent aquatic eutrophication for PBMAs, a conclusion echoed by Heller et al. and Lusk et al. Stagnari et al. and Smetana et al. [63, 64] constructed a model to analyze the economic and environmental effects of using alternative plant products over meat in the USA, revealing potential CO_2 sequestration through land use changes [65]. Similarly, Saget et al. reported a reduction in human-animal competition for land use for pea protein production in Europe and a significant reduction in global warming potential [66]. In summary, while a limited number of LCA studies have quantified the environmental impacts of meat alternatives, several methodological limitations need addressing. Key considerations include the highly processed nature of PBMAs, necessitating a balanced assessment of energy-related impacts counteracting the environmentally friendly aspects of plant-based ingredient production [59]. The building of databases for complex (multi-ingredient) food production processes and the implementation of a functional unit considering primary nutrients are essential focal points [67]. Sustainability assessments of PBMA production should encompass good agricultural practices, encompassing crop rotation, fertilizer usage, plant protection, and water utilization [68].

In the domain of meat alternatives, despite advancements in technology and attempts to formulate processed plant-based items from diverse sources, a significant hurdle in effectively substituting animal products with plant-based ingredients lies in replicating comparable sensory properties of meat. Furthermore, communication strategies regarding these novel products and individual characteristics (such as attitude and demographics) should be carefully considered during the marketing phase, particularly in countries where meat and meat-based products hold substantial significance in consumers' minds, shaped by habits, culinary traditions, and culture. Hence, both sensory and consumer science play crucial roles in comprehending how consumers perceive PBMAs, encompassing factors that drive or impede their acceptance.

As per the recent publication in *Nature Communications*, substituting meat and dairy products with plant-based alternatives could lead to additional climate and biodiversity advantages, particularly when the land freed from livestock production is reforested. This action has the potential to more than double the climate benefits and cut in half the anticipated declines in ecosystem integrity by 2050 [69]. Notably, this study is groundbreaking as it examines the global impacts of large-scale consumption of plant-based meat and milk, taking into account the intricate nature of food systems. Conducted by researchers from UVM, the International Institute for Applied Systems Analysis (IIASA), the Alliance of Bioversity International, and CIAT, the study also involved collaboration with scientists from Impossible Foods, a company specializing in plant-based meat substitutes, to ensure relevance and address gaps in existing scientific literature. Scenarios of dietary changes were developed by IIASA researchers using plant-based recipes for beef, pork, chicken, and milk provided by Impossible Foods. These recipes were crafted to be nutritionally equivalent to their animal-derived counterparts and feasible for existing food manufacturing capabilities, utilizing globally available production ingredients. It is important to note that the study was entirely under the control of the science team, and the data intentionally lacks specificity to Impossible Foods.

To maximize the environmental benefits of dietary shifts, it is crucial to restore agricultural land freed from livestock and feed production through afforestation that prioritizes biodiversity. In a 50% substitution scenario, the reduction in land-use emissions could potentially double compared to a scenario without afforestation, resulting in a total reduction of 6.3 Gt CO_2eq per year. Furthermore, this substitution scenario has the potential to decrease predicted declines in ecosystem integrity by more than half.

While the findings endorse the expanded adoption of plant-based meat alternatives, the researchers acknowledge the significant role that livestock play as a vital source of income and nutrition for smallholders in low- and middle-income countries. Livestock also hold cultural significance, contribute to risk reduction, and diversify income for smallholders. Concurrently, the looming threat of climate change poses risks to the livelihoods of these smallholder farmers. Urgent policy and management measures are imperative to mitigate environmental risks, support farmers, and bolster other stakeholders in the livestock value chain, ensuring a socially equitable and sustainable transition in the food system. This becomes especially critical given recent setbacks in achieving global food security.

The study highlights that the impacts of this transition may vary across regions due to differences in population size, dietary patterns, uneven agricultural productivity, and engagement in international trade of agricultural commodities. Notably, China experiences the main impacts on agricultural input use, while Sub-Saharan Africa and South America face significant environmental consequences. Recognizing these regional variations is crucial for designing targeted interventions that can contribute to more effective outcomes.

Food production stands out as one of the most environmentally impacted field of human activities, with agricultural practices accounting for a substantial portion of greenhouse gas emissions (GHGE), ranging from 61% to 81%, as well as 79% of acidification and 95% of eutrophication related to food impacts [70]. The challenge of nourishing a projected 10 billion people by 2050 and elevating the calorific energy supply from 30 to 45 exajoules need a shift from traditional high-calorie diets prevalent in high-income countries, which primarily consist of highly processed and animal-derived products [71, 72].

Plants continue to be the primary source of biomass utilized as a substitute for meat. Key inputs for plant-based alternatives encompass primary ingredients such as soybeans, wheat, peas, and lupine. To be considered as meat substitutes, raw grains must undergo processing to enhance nutrient availability. Studies have identified wet spinning technology as a prevalent technique for turning soy, pea, and faba beans into food-grade fibers [73]. Another potential technology for crafting textures at the nanofiber level is electrospinning, though its applications are more limited and specific, typically requiring the presence of certain substances like polyphenols or probiotics. More industrially relevant are "top-down" methods applicable to plant protein concentrates and isolates, involving low (cooking) and high moisture extrusion, proteins and hydrocolloid mixtures, and shear cell technology. It is worth noting that the latter technologies are presently mostly valid on a pilot scale only. Although processing technologies produce texturizing products, their use can differ due to resource requirements and related environmental impacts.

The primary components of the matrix in plant-based substitutes encompass cereals and pseudo cereals (e.g., chia, quinoa), legumes, and various mixtures of these. Greenhouse gas emissions from the production of main matrix components range from 0.2 to 2.1 CO_2eq. Kg^{-1} for grains (beans) and flours, 0.7 to 3.3 kg CO_2eq. Kg^{-1} for protein concentrates, and 1.8 to 13.0 kg CO_2eq. Kg^{-1} protein for isolates. Similar trends can be seen in the land usage impacts, which range from 2.0 to 5.5 m^2, 3.2 to 20.8 m^2, and 5.8 to 34.7 m^2 for isolates, concentrates, and raw materials, respectively. Furthermore, when considering meat substitutes, it is essential to note that extensive processing and the addition of minor components like spices and preservatives usually add an extra 13–26% to the resource demand, consequently elevating the environmental effect of plant-based meat [74–76].

Plant-based foods in human diets exhibit greenhouse gas emissions (GHGE) that are half as much as those of animal-based foods, with values of 4963 $TgCO_2$eq and 9923 $TgCO_2$eq, respectively [77]. Additionally, when considering GHGE on a protein basis, proteins based on animal show significantly higher values compared to proteins found in plant-based meat substitutes. A varying degree of GHG emission ranging from 34% in fish to high of 93% in beef herd has been found. This suggests that, in general, plant-based proteins contribute to lower environmental impacts when used in meat alternates compared to various types of meat. However, it is crucial to note that processed plant-based meat substitutes can have 1.6–7 times higher environmental impacts than that of less processed plant protein sources like tofu, pulses, and peas.

Research conducted within the scope of the Protein2Food project by Detzel et al. revealed that under some circumstances, extruded plant-based meat substitutes could have a carbon footprint similar to that of chicken meat and may even exceed it in terms of resource demand (land, energy, and water) [78]. Both animal- and plant-based ingredients exhibit a wide variation in impacts, with overlapping results in the range of impact assessments, making it challenging to establish a universal base case for evaluation in all instances. While beef is generally recognized as a product to have a significant environmental effect, surpassing majority of meat substitute ingredients, some protein sources like microalgae show that based on weight, their GHGE and NRE demand can exceed those of beef and other plant raw materials. Cell-based cultures and insects have a tendency to have a greater environmental impact when used as meat substitutes. In protein comparisons, for the majority of categories, a range from most significant to least impacting can be found (excluding water footprint): beef, microalgae, cell meat, poultry meat, insects, and plants.

A recent publication in *Future Foods* argues that plant-based meat and dairy alternatives, designed to replicate the taste, texture, and overall eating experience of animal products, offer a more effective means of reducing the demand for meat and dairy compared to merely encouraging the consumption of vegetarian whole foods. Conducted by psychologists at the University of Bath, the study concludes that these alternatives present a healthier and environmentally sustainable solution that takes consumer preferences and behavior into account. The review, encompassing 43 studies examining the health and environmental impacts of plant-based foods, along with consumer attitudes, revealed several noteworthy findings. Nearly 90% of consumers who incorporated plant-based meat and dairy into their diet were meat-eaters or flexitarians, emphasizing the potential appeal of these alternatives beyond traditional vegetarian demographics. Products with a taste, texture, and price resembling processed meat were identified as having the highest potential for replacing traditional meat.

Furthermore, the study observed lower greenhouse gas emissions associated with plant-based products compared to the animal products they aim to replace. Using pea protein as a substitute for a small portion of German beef, for example, could cut emissions by as much as 8 million tons per year. Plant-based burgers, in comparison to beef burgers, were found to be linked with up to 98% fewer greenhouse gas emissions.

The authors of the paper suggest that plant-based products generally require less agricultural land and water and cause less pollution than their animal counterparts. Health-focused studies also indicated that plant-based alternatives often have superior nutritional profiles compared to animal products. It is worth noting that 40% of traditional meat products were classified as "less healthy" compared to only 14% of plant-based alternatives according to the UK's Nutrients Profiling Model. Additional benefits noted in the research include the positive effects of plant-based meat and dairy on weight loss, muscle mass building, and potential use in addressing specific health conditions. The study suggests that future advancements in processing and ingredient innovation may lead to further nutritional enhancements in plant-based foods. Dr. Chris Bryant, one of the report authors from the University of Bath, emphasized the overwhelming evidence supporting the sustainability and health benefits of plant-based alternatives, highlighting their ability to cater to consumers' essential preferences for taste, price, and convenience.

Sustainability principles and practices are increasingly crucial for both managers and policymakers. The adoption of more sustainable business approaches, particularly in supply chain management, has substantial implications. Research suggests that the effectiveness of sustainable supply chain management strategies varies across industries [79]. The industry-specific nature of these approaches poses unique challenges in integrating sustainability practices, emphasizing the need to address sustainability issues with consideration for the specific characteristics of each industry. The food industry stands out as a significant example of a supply chain where the shift towards sustainability and alterations in management approaches are crucial to address the impacts of climate change. As noted by Yakovleva, "the food sector is constantly innovating with alternative strategies to decrease burdens on the natural environment and improve social and ethical issues in supply chain through various production, marketing, labelling, accreditation schemes and initiatives" [80]. Nevertheless, there exists considerable diversity within the food sector regarding the factors driving sustainable innovation and sustainability-focused managerial methods. This diversity is evident due to variations in food products, such as price volatility, levels of vertical integration, mechanisms of price transmission, and reliance on imports in national trade balances.

Among various food products, the scholarly and practical interest in meat production is notable due to its environmental implications. This impact has prompted an increased emphasis on researching the meat supply chain, with numerous studies dedicated to its analysis. For example, scholars like Leroy and Praet underscore the significance of meat production traditions and their cultural role within food production systems [81]. Other researchers delve into the institutional environment's role in enhancing supply chain efficiency [82], as well as the diverse array of products resulting from different stages in the meat production process. Additionally, some studies concentrate on consumer behavior and awareness, specifically exploring the repercussions of the growing perception of meat products as having a substantial impact on climate change.

The Covid-19 pandemic has had a significant impact on consumer behavior in a number of ways: It has increased consumer price sensitivity; it has accelerated online shopping across every category; it has caused consumers to pay more attention to health, wellness, and hygiene. There has been a significant shift in consumer attitudes towards sustainable packaging. Prior to Covid-19, there was a significant increase in public awareness of packaging leakage into the environment. FMCG companies and retailers, as well as regulators, were making significant pledges to sustainable packaging.

Mintel's research [83] also shows a link between environmental concerns and the rise of meat alternatives and vegan products. Eating more plant-based meals is viewed as an excellent strategy to reduce human effect on the environment by 48% of British consumers. Additionally, 75% of flexitarian customers would choose one meatless food product over another if it had environmentally friendly packaging.

In the vegan meats market, there is a huge opportunity for brands to differentiate themselves through more sustainable packaging. Packaging that is recyclable and has a minimal carbon footprint will appeal to consumers who care about the environment, making firms stand out as leaders in sustainability within their industry.

The quality of meat production commences with meticulous animal care, beginning from prenatal stages through to feeding during its life, contributing to the development of a proper carcass, fat deposition, and the quantity and size of muscular fibers. The overarching concept of high-quality meat products is rooted in a combination of various factors that can be examined individually or correlated. Industry relies on key parameters, including (1) visual quality (in terms of color, muscle, and fat), (2) gustatory quality (covering texture, flavor, and odor), (3) nutritional quality, and (4) food safety. The ultimate quality of the end product is influenced by the proper treatment of animals, with a focus on their diet and living conditions, followed by appropriate processing facilities and controls, packaging, transportation, and storage. Different countries implement various regulations and norms as best practices to enhance overall food quality and safety for consumers.

During storage, meat products undergo deterioration due to pigment oxidation, oxidative rancidity, microbial growth, and surface dehydration. Color, a crucial attribute evaluated by consumers during the purchasing process, is determined by the state of the meat pigment, myoglobin, present on the meat surface. The transition of fresh meat color to red, also known as "blooming," is a result of the presence of oxymyoglobin, the oxygenated state of myoglobin.

Moving towards plant-based meats is considered by many consumers as a stride toward adopting a sustainable lifestyle and contributing to a healthier future for the planet. Nielsen reports that 61% of consumers who are concerned about the environmental impact of livestock on climate change would decrease their meat consumption, with 43% opting to replace animal-based protein with plant-based alternatives. Environmentally conscious consumers often express worries about the sustainability of packaging.

The use of flexible packaging with compatible closures serves to underscore your brand's commitment to the environment. The lightweight nature of flexible packaging results in a smaller carbon footprint compared to rigid packaging, as it reduces fossil fuel and water usage while lowering greenhouse gas emissions. Moreover, flexible packaging contributes to a reduction in transportation needs, materials sent to landfills, and overall food waste. Consumers who purchase multi-serving packages can conveniently reseal them for future use, minimizing food wastage.

The evolution of flexible packaging, along with the development of mono-material films and compatible closures, provides fully recyclable alternatives for various commercial markets, including dried goods like cereals and baking ingredients.

References

1. Poore J, Nemecek T. Reducing food's environmental impacts through producers and consumers. Science. 2018;360(6392):987–92.
2. Bodirsky BL, et al. The ongoing nutrition transition thwarts long-term targets for food security, public health and environmental protection. Sci Rep. 2020;10(1):1–12.
3. Clark MA, et al. Global food system emissions could preclude achieving the 1.5° and 2°C climate change targets. Science. 2018;360(6390):987–92.
4. Uwizeye A, et al. Livestock production and greenhouse gas emissions: impact of climate change and mitigation strategies. Environ Res Lett. 2020;15(3):035005.
5. Harwatt H. Including animal to plant protein shifts in climate change mitigation policy: a proposed three-step strategy. Clim Pol. 2019;19(4):533–41.
6. Willett W, et al. Food in the Anthropocene: the EAT–lancet commission on healthy diets from sustainable food systems. Lancet. 2019;393(10170):447–92.
7. González AD, et al. The protein content of animal products: a review. Qual Assur Saf Crops Foods. 2020;12(2):125–38.
8. Xu X, et al. Global and regional food demand of crop and livestock. Environ Res Lett. 2021;16(3):035009.
9. Wowra V, et al. Evaluation of nutrient management strategies in a coupled human-natural system: a case study of nitrogen and phosphorus in the Weser-ems region, Germany. Sci Total Environ. 2021;784:147040.

10. Cheah I, Shimul AS, Liang J, Phau I. Drivers and barriers toward reducing meat consumption. Appetite. 2020;149:104636. https://doi.org/10.1016/j.appet.2020.104636.
11. Milford AB, Le Mouël C, Bodirsky BL, Rolinski S. Drivers of meat consumption. Appetite. 2019;141:104313. https://doi.org/10.1016/j.appet.2019.06.005.
12. Smet D, Stefaan, and Els Vossen. Meat: the balance between nutrition and health. A review. Meat Sci. 2016;120:145–56. https://doi.org/10.1016/j.meatsci.2016.04.008.
13. Hyland J, Henchion M, McCarthy MB, McCarthy S. The role of meat in strategies to achieve a sustainable diet lower in greenhouse gas emissions: a review. Meat Sci. 2017;132:189–95. https://doi.org/10.1016/j.meatsci.2017.04.014.
14. Bonnet C, Bouamra-Mechemache Z, Réquillart V, Treich N. Viewpoint: regulating meat consumption to improve health, the environment and animal welfare. Food Policy. 2020;97:101847. https://doi.org/10.1016/j.foodpol.2020.101847.
15. Fresán U, Sabaté J. Vegetarian diets: planetary health and its alignment with human health. Adv Nutr. 2019;10:S380–8. https://doi.org/10.1093/advances/nmz019.
16. Rosi A, Mena P, Pellegrini N, Turroni S, Neviani E, Ferrocino I, Di Cagno R, et al. Environmental impact of omnivorous, Ovo-lacto-vegetarian, and vegan diet. Sci Rep. 2017;7(1) https://doi.org/10.1038/s41598-017-06466-8.
17. Kemper JA. Motivations, barriers, and strategies for meat reduction at different family lifecycle stages. Appetite. 2020;150:104644. https://doi.org/10.1016/j.appet.2020.104644.
18. Bohrer BM. Review: nutrient density and nutritional value of meat products and non-meat foods high in protein. Trends Food Sci Technol. 2017;65:103–12. https://doi.org/10.1016/j.tifs.2017.04.016.
19. Domingo JL, Nadal M. Carcinogenicity of consumption of red meat and processed meat: a review of scientific news since the IARC decision. Food Chem Toxicol. 2017;105:256–61. https://doi.org/10.1016/j.fct.2017.04.028.
20. Lippi G, Mattiuzzi C, Cervellin G. Meat consumption and cancer risk: a critical review of published meta-analyses. Crit Rev Oncol Hematol. 2016;97:1–14. https://doi.org/10.1016/j.critrevonc.2015.11.008.
21. Europe PMC. Europe PMC. https://europepmc.org/article/med/31525258 (n.d.).
22. Sasso A, Latella G. Dietary components that counteract the increased risk of colorectal cancer related to red meat consumption. Int J Food Sci Nutr. 2017;69(5):536–48. https://doi.org/10.1080/09637486.2017.1393503.
23. IPCC. Climate change 2007: synthesis report. Contribution of working groups I, II and III to the fourth assessment report of the intergovernmental panel on climate change: university of southern Queensland repository. https://research.usq.edu.au/item/q6wy0/climate-change-2007-synthesis-report-contribution-of-working-groups-i-ii-and-iii-to-the-fourth-assessment-report-of-the-intergovernmental-panel-on-climate-change (n.d.).
24. Poore J, Nemecek T. Reducing food's environmental impacts through producers and consumers. Science. 2018;360(6392):987–92. https://doi.org/10.1126/science.aaq0216.
25. Xu X, Sharma P, Shu S, Lin T-S, Ciais P, Tubiello FN, Smith P, Campbell N, Jain AK. Global greenhouse gas emissions from animal-based foods are twice those of plant-based foods. Nat Food. 2021;2(9):724–32. https://doi.org/10.1038/s43016-021-00358-x.
26. Aiking H, De Boer J. The next protein transition. Trends Food Sci Technol. 2020;105:515–22. https://doi.org/10.1016/j.tifs.2018.07.008.
27. Ritchie H. Land use. Our world in data, February 16, 2024. https://ourworldindata.org/land-use.
28. FAOHome. Home | Food and Agriculture Organization of the United Nations. http://www.fao.org/home/en (n.d.).
29. Whitton C, Bogueva D, Marinová D, Phillips CJC. Are we approaching peak meat consumption? Analysis of meat consumption from 2000 to 2019 in 35 countries and its relationship to gross domestic product. Animals. 2021;11(12):3466. https://doi.org/10.3390/ani11123466.
30. Barnard ND, Levin S, Trapp C. Meat consumption as a risk factor for type 2 diabetes. Nutrients. 2014;6(2):897–910. https://doi.org/10.3390/nu6020897.

31. Godfray HC, et al. Meat consumption, health, and the environment. Science. 2018;361(6399) https://doi.org/10.1126/science.aam5324.
32. Afshin A, Sur PJ, Fay K, Cornaby L, Ferrara G, Salama J, Mullany EC, et al. Health effects of dietary risks in 195 countries, 1990–2017: a systematic analysis for the global burden of disease study 2017. Lancet. 2019;393(10184):1958–72. https://doi.org/10.1016/s0140-6736(19)30041-8.
33. Google Books. Sustainable healthy diets. https://books.google.com/books?hl=en&lr=&id=pH-4DwAAQBAJ&oi=fnd&pg=PP2&ots=JrkfqrcDfi&sig=n31jhS9DiXZSP9W0BIvsygBLtOM (n.d.).
34. Pereira P, Vicente F. Meat nutritional composition and nutritive role in the human diet. Meat Sci. 2013;93(3):586–92. https://doi.org/10.1016/j.meatsci.2012.09.018.
35. Cole E, Goeler-Slough N, Cox AN, Nolden AA. Examination of the nutritional composition of alternative beef burgers available in the United States. Int J Food Sci Nutr. 2021;73(4):425–32. https://doi.org/10.1080/09637486.2021.2010035.
36. Cutroneo S, Angelino D, Tedeschi T, Pellegrini N, Martini D. Nutritional quality of meat analogues: results from the food labelling of Italian products (FLIP) project. Front Nutr. 2022;9 https://doi.org/10.3389/fnut.2022.852831.
37. Bryngelsson S, Moshtaghian H, Bianchi M, Hallström E. Nutritional assessment of plant-based meat analogues on the Swedish market. Int J Food Sci Nutr. 2022;73(7):889–901. https://doi.org/10.1080/09637486.2022.2078286.
38. De Marchi M, Costa A, Pozza M, Goi A, Manuelian CL. Detailed characterization of plant-based burgers. Sci Rep. 2021;11(1) https://doi.org/10.1038/s41598-021-81684-9.
39. Gorissen SHM, Crombag JJR, Senden JM, et al. Protein content and amino acid composition of commercially available plant-based protein isolates. Amino Acids. 2018;50(12):1685–95. https://doi.org/10.1007/s00726-018-2640-5.
40. Hunt J. Bioavailability of iron, zinc, and other trace minerals from vegetarian diets. Am J Clin Nutr. 2003;78(3):633S–9S. https://doi.org/10.1093/ajcn/78.3.633s.
41. Kumar M, Tomar M, Punia S, Dhakane-Lad J, Dhumal S, Changan S, Senapathy M, et al. Plant-based proteins and their multifaceted industrial applications. LWT. 2022;154:112620. https://doi.org/10.1016/j.lwt.2021.112620.
42. Zahari I, Östbring K, Purhagen J, Rayner M. Plant-based meat analogues from alternative protein: a systematic literature review. Foods. 2022;11(18):2870. https://doi.org/10.3390/foods11182870.
43. De Angelis D, Kaleda A, Pasqualone A, Vaikma H, Tamm M, Tammik ML, Squeo G. Physicochemical and sensorial evaluation of meat analogues produced from dry-fractionated pea and oat proteins. Foods. 2020;9(12):1754. https://doi.org/10.3390/foods9121754.
44. Wittek P, Ellwanger F, Karbstein HP, Emi MA. Morphology development and flow characteristics during high moisture extrusion of a plant-based meat analogue. Foods. 2021;10(8):1753. https://doi.org/10.3390/foods10081753.
45. Sha L, Xiong YL. Plant protein-based alternatives of reconstructed meat: science, technology, and challenges. Trends Food Sci Technol. 2020;102:51–61. https://doi.org/10.1016/j.tifs.2020.05.022.
46. Dagevos H. Exploring flexitarianism. In: Practice, progress, and proficiency in sustainability; 2016. p. 233–43. https://doi.org/10.4018/978-1-4666-9553-5.ch012.
47. Nezlek JB, Forestell CA. Meat substitutes: current status, potential benefits, and remaining challenges. Curr Opin Food Sci. 2022;47:100890. https://doi.org/10.1016/j.cofs.2022.100890.
48. Nezlek JB, Forestell CA. The carbon footprint of dairy production systems through partial life cycle assessment. J Dairy Sci. 2010;93(3):1266–82. https://doi.org/10.3168/jds.2009-2162.
49. Google Books. Livestock's long shadow. https://books.google.com/books?hl=en&lr=&id=1B9LQQkm_qMC&oi=fnd&pg=PR16&ots=LQU2kTaGmM&sig=oAot0PpWiN-xYy7Vnv-0pU5ljAI (n.d.).
50. Oecd F. OECD-FAO agricultural outlook 2022–2031. Policy commons, August 29, 2022. https://policycommons.net/artifacts/2652558/oecd-fao-agricultural-outlook-2022-2031/3675435/.

51. Aiking H. Future protein supply. Trends Food Sci Technol. 2011;22(2–3):112–20. https://doi.org/10.1016/j.tifs.2010.04.005.
52. Tukker A, Alexandra Goldbohm R, De Koning A, Verheijden MW, Kleijn R, Oliver W, Pérez-Domínguez I, Rueda-Cantuche JM. Environmental impacts of changes to healthier diets in Europe. Ecol Econ. 2011;70(10):1776–88. https://doi.org/10.1016/j.ecolecon.2011.05.001.
53. ISO. ISO 14040:2006. https://www.iso.org/standard/37456.html (n.d.).
54. ISO. ISO 14044:2006. https://www.iso.org/standard/38498.html (n.d.).
55. Froldi F, Lamastra L, Trevisan M, Mambretti D, Moschini M. Environmental impacts of cow's milk in northern Italy: effects of farming performance. J Clean Prod. 2022;363:132600. https://doi.org/10.1016/j.jclepro.2022.132600.
56. Bryant C. Plant-based animal product alternatives are healthier and more environmentally sustainable than animal products. Future Foods. 2022;6:100174. https://doi.org/10.1016/j.fufo.2022.100174.
57. Detzel A, Krüger M, Busch M, Ferrer J, Varela C, Manners R, Bez J, Zannini E. Life cycle assessment of animal-based foods and plant-based protein-rich alternatives: an environmental perspective. J Sci Food Agric. 2021;102(12):5098–110. https://doi.org/10.1002/jsfa.11417.
58. Pulvento C, Riccardi M, Lavini A, D'Andria R, Ragab R. Parameterization and field validation of Saltmed model for grain Amaranth tested in South Italy. Irrig Drain. 2015;64(1):59–68. https://doi.org/10.1002/ird.1905.
59. Smetana S, Mathys A, Knoch A, Heinz V. Meat alternatives: life cycle assessment of most known meat substitutes. Int J Life Cycle Assess. 2015;20(9):1254–67. https://doi.org/10.1007/s11367-015-0931-6.
60. Mejía MA, Fresán U, Harwatt H, Oda K, Uriegas-Mejia G, Sabaté J. Life cycle assessment of the production of a large variety of meat analogs by three diverse factories. J Hunger Environ Nutr. 2019;15(5):699–711. https://doi.org/10.1080/19320248.2019.1595251.
61. Goldstein B, Moses R, Sammons N, Birkved M. Potential to curb the environmental burdens of American beef consumption using a novel plant-based beef substitute. PLoS One. 2017;12(12):e0189029. https://doi.org/10.1371/journal.pone.0189029.
62. Goldstein B, Moses R, Sammons N, Birkved M. Meat consumption, health, and the environment. Science. 2018;361(6399) https://doi.org/10.1126/science.aam5324.
63. Stagnari F, Maggio A, Galieni A, Pisante M. Multiple benefits of legumes for agriculture sustainability: an overview. Chem Biol Technol Agric. 2017;4(1) https://doi.org/10.1186/s40538-016-0085-1.
64. Smetana S, Profeta A, Voigt R, Kircher C, Heinz V. Meat substitution in burgers: nutritional scoring, sensorial testing, and life cycle assessment. Future Foods. 2021;4:100042. https://doi.org/10.1016/j.fufo.2021.100042.
65. Center for Sustainable Systems. Beyond meat's beyond burger life cycle assessment: a detailed comparison between a plant-based and an animal-based protein source. https://css.umich.edu/publications/research-publications/beyond-meats-beyond-burger-life-cycle-assessment-detailed (n.d.).
66. Lusk JL, Blaustein-Rejto D, Shah S, Tonsor GT. Impact of plant-based meat alternatives on cattle inventories and greenhouse gas emissions. Environ Res Lett. 2022; https://doi.org/10.1088/1748-9326/ac4fda.
67. Saget S, Costa MP, Barilli E, Vasconcelos MW, Santos CS, Styles D, Williams M. Substituting wheat with chickpea flour in pasta production delivers more nutrition at a lower environmental cost. Sustain Prod Consumtion. 2020;24:26–38. https://doi.org/10.1016/j.spc.2020.06.012.
68. Bager S, Persson M, Dos Reis TNP. Eighty-six EU policy options for reducing imported deforestation. One Earth. 2021;4(2):289–306. https://doi.org/10.1016/j.oneear.2021.01.011.
69. Google Books. Integration of environment and nutrition in life cycle assessment of food items: opportunities and challenges. https://books.google.com/books?hl=en&lr=&id=uqJfEAAAQBAJ&oi=fnd&pg=PR10&ots=AoMAob46yX&sig=AmnQnpfm6Um9G_vlVRNh3l5X-aVo (n.d.).

70. Google Books. Reducing food's environmental impacts through producers and consumers. Science. 2018;360(6392):987–92. https://doi.org/10.1126/science.aaq0216.
71. Bodirsky BL, Dietrich JP, Martinelli E, Stenstad A, Pradhan P, Gabrysch S, Mishra A, et al. The ongoing nutrition transition thwarts long-term targets for food security, public health and environmental protection. Sci Rep. 2020;10(1) https://doi.org/10.1038/s41598-020-75213-3.
72. Heusala H, Sinkko T, Mogensen L, Knudsen MT. Motivations, barriers, and strategies for meat reduction at different family lifecycle stages. Appetite. 2020;150:104644. https://doi.org/10.1016/j.appet.2020.104644.
73. Heusala H, Sinkko T, Mogensen L, Knudsen MT. Alternative protein sources as techno-functional food ingredients. Annu Rev Food Sci Technol. 2021;12(1):93–117. https://doi.org/10.1146/annurev-food-062520-093642.
74. Heusala H, Sinkko T, Sözer N, Hytönen E, Mogensen L, Knudsen MT. Carbon footprint and land use of oat and Faba bean protein concentrates using a life cycle assessment approach. J Clean Prod. 2020;242:118376. https://doi.org/10.1016/j.jclepro.2019.118376.
75. Saerens W, Smetana S, Van Campenhout L, Lammers V, Heinz V. Life cycle assessment of burger patties produced with extruded meat substitutes. J Clean Prod. 2021;306:127177. https://doi.org/10.1016/j.jclepro.2021.127177.
76. Saerens W, Smetana S, Van Campenhout L, Lammers V, Heinz V. Meat substitution in burgers: nutritional scoring, sensorial testing, and life cycle assessment. Future Foods. 2021;4:100042. https://doi.org/10.1016/j.fufo.2021.100042.
77. Saerens W, Smetana S, Van Campenhout L, Lammers V, Heinz V. Global greenhouse gas emissions from animal-based foods are twice those of plant-based foods. Nature Food. 2021;2(9):724–32. https://doi.org/10.1038/s43016-021-00358-x.
78. Saerens W, Smetana S, Van Campenhout L, Lammers V, Heinz V. Life cycle assessment of animal-based foods and plant-based protein-rich alternatives: an environmental perspective. J Sci Food Agric. 2021;102(12):5098–110. https://doi.org/10.1002/jsfa.11417.
79. Cagno E, Negri M, Neri A, Giambone M. One framework to rule them all: an integrated, multi-level and scalable performance measurement framework of sustainability, circular economy and industrial Symbiosis. Sustain Prod Consum. 2023;35:55–71. https://doi.org/10.1016/j.spc.2022.10.016.
80. Yakovleva N, Sarkis J, Sloan TW. Sustainable benchmarking of supply chains: the case of the food industry. Int J Prod Res. 2012;50(5):1297–317. https://doi.org/10.1080/00207543.2011.571926.
81. Hübel C, Schaltegger S. Barriers to a sustainability transformation of meat production practices – an industry actor perspective. Sustain Prod Consum. 2022;29:128–40. https://doi.org/10.1016/j.spc.2021.10.004.
82. Leroy F, Praet I. Meat traditions. The co-evolution of humans and meat. Appetite. 2015;90:200–11. https://doi.org/10.1016/j.appet.2015.03.014.
83. Mintel. Plant-based push: UK sales of meat-free foods shoot up. https://www.mintel.com/press-centre/plant-based-push-uk-sales-of-meat-free-foods-shoot-up-40-between-201419/#:~:text=Meanwhile%2C%20sales%20of%20meat%2Dfree,%C2%A31.1%20billion%20by%202024 (n.d.).

Chapter 4
Nutritional Composition and Health Implication of Meat Substitutes

Plant-based meat analogues have attracted considerable attention within the food industry as potential substitutes for traditional meat products. These alternatives have the capacity to replicate the fibrous texture, taste, and nutritional attributes of authentic meat while yielding positive consequences for environmental sustainability, human health, and animal welfare. The fundamental constituents utilized in the creation of plant-based meat analogs are plant proteins, serving as the primary structural components for generating a meat-like fibrous architecture through the deconstruction and reassembly of proteins. Furthermore, non-animal ingredients such as oil, pigment, and adhesives are incorporated into the process to shape, color, and flavor the analogues, closely resembling real animal meat.

Various techniques, including electrospinning, 3D printing, shearing, and high-moisture extrusion, are employed in the processing of plant-based meat analogues to achieve a fiber-like structure akin to conventional meat. Notably, Impossible Foods™ has genetically modified Pichia pastoris to produce soy leghemoglobin, an additive imparting a fresh meat color to their plant-based meat analogs. Post-cooking, the analogues exhibit browning and emit meat-like aromas. Concerning flavor, the primary raw materials, namely, legume proteins, contribute a beany taste that may be undesirable and impact the overall product flavor. To mitigate this beany flavor, natural spices, along with meaty and savory aromas, are introduced to mask and substitute it with a more palatable meat flavor.

The preference for plant-based meat analogues among consumers is influenced not only by their appearance, texture, and flavor but also by their nutritional content. Bohrer [1] conducted an investigation into the nutritional composition of meat analog products as compared to traditional meat products. The findings revealed that meat analog products exhibited lower levels of saturated fatty acids and cholesterol, along with higher quantities of carbohydrates and dietary fiber in comparison to traditional meat products. In a separate study, Zhou et al. [2] examined the in vitro digestion characteristics of beef analogues in contrast to beef. The results indicated

S. Ray, S. Suri, *The Sustainable Plate: Unveiling the Science of Meat Substitutes and Impact on Global Health*, https://doi.org/10.1007/978-3-031-94604-2_4

that the digestibility of protein and fat in beef analogs was lower, although the protein in beef analogs demonstrated a faster digestion rate in the stomach.

Xie et al.'s study [3] focused on plant-based beef and pork analogues, revealing lower digestibility and a reduced release of potential bioactive peptides compared to beef and pork. Given that previous investigations primarily concentrated on analyzing the nutritional disparities between plant and animal meat, a more comprehensive examination of their nutritional and digestive attributes would be advantageous for refining plant meat formulations and creating health-conscious products.

The prevalence of vegetarianism and veganism is on the rise in Spain. According to data presented by the Lantern consultancy in 2021, there was a 34% increase in the number of individuals adopting a vegetarian or vegan diet from 2019 to 2021, surpassing the 27% growth observed in the preceding two-year period [4]. This shift is primarily propelled by ethical considerations regarding animal welfare and environmental concerns, coupled with the perception that such diets offer superior health benefits compared to those incorporating animal-derived products. Concurrently, the market has witnessed a surge in the availability of meat analogues.

Tofu, a soy-based product commonly utilized as a meat substitute, originated in China. Notably, the first patent for soy protein textured to emulate meat was granted in the United States in 1955 [5]. Although the consumption of tofu and soy protein was limited in Western countries until 1960, it has since demonstrated sustained growth. Between 2015 and 2021, a notable 4400 plant-based products were introduced to the market as substitutes for various animal-derived products. The global meat analogue market is presently dominated by European nations, particularly the United Kingdom, Germany, Italy, France, Sweden, Belgium, Norway, and the Netherlands, which are leading the way in innovations within the meat protein alternatives sector.

Conducted by ProVeg International, a study reveals a notable 20% increase in the sales volume of plant-based products designed to emulate foods of animal origin in Spain over the past 2 years. Within this market, plant-based alternatives such as "milks," "meat," and "yogurt" have exhibited substantial prominence [6]. The report further underscores the receptiveness of Spanish consumers towards incorporating plant-based meat substitutes into their regular purchases, surpassing the willingness observed in other countries. Notably, 47% of Spanish consumers express a readiness to acquire these products if they closely replicate the flavor and texture of their animal-derived counterparts [7].

From a health perspective, adopting a dietary pattern characterized by reduced consumption of animal-origin products, particularly red meat, and substituting them with plant-derived foods has been associated with lower overall mortality in epidemiological studies [8–10]. However, it is imperative to recognize that the health benefits associated with a plant-based diet cannot be seamlessly extrapolated to the replacement of meat products with plant-based analogues. The limited understanding of the composition of these substitutes, marked by considerable formulation variability, is a noteworthy challenge. Frequently derived from vegetable protein concentrates and isolates, the inclusion of other essential plant components such as vitamin C, carotenoids, and polyphenols cannot be guaranteed.

Plant-based ingredients serve as a rich source of vitamins, minerals, and antioxidants, all of which confer various health benefits. Substantiated by research [11], plant-based diets exhibit potential advantages in weight management, diabetes regulation, and the enhancement of the microbiome. Furthermore, these dietary patterns may contribute to the prevention of health conditions such as hypertension and cardiovascular disease [12]. Official health organizations, including the World Health Organization (WHO) [13], express concerns about carcinogens present in red and processed meats, highlighting the potential links to increased cancer risk. In alignment with this, the American Heart Association (AHA) [14] recommends the substitution of plant protein for meat as a strategy to improve heart health.

However, it is imperative to acknowledge that certain plant-based products may incorporate fillers, added sodium, and elevated levels of saturated fats. Additionally, variability exists in the fortification of essential nutrients, such as vitamin B12, among plant-based meat manufacturers, rendering some products deficient in comparison to their counterparts. Furthermore, plant-based meats may exhibit lower concentrations of zinc and other minerals in comparison to conventional meat.

In a study conducted in 2020 [15], a cohort of 36 participants underwent an 8-week dietary intervention, during which they incorporated meat products into their routine diet. Subsequently, they transitioned to a plant-based diet for another 8 weeks, maintaining consistency in the remaining aspects of their dietary patterns.

Upon completion of the study, notable enhancements in various cardiovascular disease risk factors were observed among the participants. Specifically, there was a reduction in low-density lipoprotein cholesterol levels. Throughout the plant-based phase, participants exhibited a decreased intake of saturated fat and an increased consumption of dietary fiber compared to the meat-based phase. It is noteworthy that sodium intake remained relatively consistent throughout the entire duration of the study.

The escalating consumer preference for plant-based diets is driving advancements in the development of plant-based meat alternatives (PBMAs), typically composed of plant proteins and other constituents. The food industry is diversifying its PBMA product offerings, aiming to replicate the visual, textural, oral, and gustatory attributes of conventional meat products. Consumers are increasingly embracing these alternatives due to concerns related to the health, environmental, and ethical implications associated with traditional meat consumption. Current meat consumption patterns have been linked to adverse environmental consequences, including greenhouse gas emissions, pollution, and biodiversity loss, as well as detrimental effects on human health, such as cardiovascular disease, cancer, and obesity. Transitioning towards a more plant-based diet is seen as a potential solution to mitigate or eliminate these issues. PBMAs are positioned as facilitators of the shift away from animal-based products, contributing to the establishment of a contemporary food system that promotes the well-being of humans, animals, and the environment.

The adoption of a plant-centric diet, particularly in affluent nations representing only 16 percent of the global population, holds the promise of reducing greenhouse gas emissions by approximately 61 percent and enhancing carbon sequestration

[16]. The meat substitute sector significantly contributes to the growth of the plant-based market, with global alternative protein consumption reaching around 13 million metric tons in 2020 [17]. It is anticipated that the PBMA product category would display unique functional, nutritional, physical, and sensory attributes.

Achieving this on a large scale and cost-effectively requires the food industry to adeptly discern appropriate ingredient combinations and manufacturing processes aligned with market trends. Projections for the global plant-based foods market, primarily encompassing PBMAs and beverages, anticipate an estimated value of USD 162 billion by 2030, indicating substantial growth compared to the USD 4.6 billion in 2018 and the USD 29.4 billion in 2020 [18]. Despite the relatively small sector size and the current challenge of consumer demand surpassing industry supply chain capabilities, these robust growth figures underscore a promising long-term outlook for investments in alternative proteins.

The increasing integration of plant-based meat alternatives (PBMAs) into the food supply chain poses potential new challenges related to food safety and nutrition, which could give rise to novel health concerns. Consequently, a comprehensive understanding of the molecular and physicochemical characteristics of plant-derived ingredients becomes imperative. It is crucial for the emerging plant-based food industry to meticulously formulate its products, taking into account considerations of food safety, nutrition, and overall healthiness. The resolution of these issues is contingent upon the specific nature of ingredients and processes utilized in PBMA production, emphasizing the need for control over nutrient composition, digestibility, and bio accessibility. Addressing these concerns should be approached on a case-by-case basis.

Researchers emphasize the significance of furnishing consumers with information about the nutritional quality of PBMAs compared to traditional meat. This provision of information aims to empower consumers to make more informed decisions regarding their dietary choices [19].

The recently introduced plant-based meat alternatives (PBMAs) are intended to function as substitutes for traditional meat products. Their nutritional composition, health implications, and safety considerations are contingent upon the specific ingredients incorporated and the type of processing operations employed during their manufacture [20]. Due to their distinct compositions and structures compared to authentic meat products, PBMAs are likely to yield divergent effects on nutrition, health, and safety impacts.

A recent surge has been witnessed in the development of next-generation plant-based meat alternatives, primarily targeting omnivores who seek the sensory attributes of meat. Currently, PBMA products predominantly manifest as burger patties, mince, sausages, and nuggets. However, there is a growing interest in the creation of more intricate and complex products, such as analogues resembling chicken breast, beefsteak, or pork chops. Environmental sustainability is a key consideration in the engineering of PBMAs, with an emphasis on faithfully replicating the appealing appearance, texture, and taste of authentic meat products. While some PBMAs closely mimic the nutritional composition of real meat, others may prioritize different aspects. Notably, the majority of PBMAs fall into the category of ultra-processed foods, potentially leading to variations in digestion and absorption rates compared to authentic meat, thereby influencing their nutritional and health effects [21]. Multiple challenges exist in the domains of food safety and nutrition within the realm of plant-based foods. These challenges encompass diverse aspects, including the presence of different types of chemical and microbial contaminants in ingredients, concerns about food adulteration, elevated levels of food additives, the use of genetically modified components, mislabeling, the introduction of new allergens, potential vitamin or mineral deficiencies, and alterations in macronutrient composition (such as protein, carbohydrate, or fat content). Therefore, careful consideration of these issues is of utmost importance in the development of the next generation of plant-based foods.

Strong arguments are mounting to support a shift from industrial animal husbandry to benefit environment, animal welfare, human health, and public health. Plant-based animal product substitutes, or PB-APAs, are a very practical way to reduce the consumption of animal products since they successfully meet the three main factors influencing consumer decisions: flavor, cost, and convenience. Concerning the sustainability of the environment, PB-APAs exhibit greater sustainability compared to animal products across various parameters, encompassing greenhouse gas emissions, water utilization, land occupation, and other ecological indicators. From a health perspective, PB-APAs offer numerous advantages, including generally favorable nutritional profiles, helping with weight loss and muscular growth, and suitability for specific health conditions. Furthermore, several studies suggest strategies for enhancing the healthiness of PB-APAs through the incorporation of optimal ingredients and improving processing techniques. Consumers and policymakers must reject naturalistic heuristics regarding PB-APAs and instead recognize their numerous advantages for the environment, public health, individual health, and animal welfare as traditional meat producers increasingly shift to plant-based meat products.

Primarily, there is a compelling environmental argument against animal agriculture. Various modeling approaches, life cycle analyses, and land use evaluations consistently indicate that diets with a high proportion of animal-based foods have more adverse environmental impacts than plant-based diets, as demonstrated by systematic reviews. In addition to direct greenhouse gas emissions, animal agriculture contributes to deforestation, freshwater depletion, and eutrophication. Reducing meat intake has been recommended in prominent papers, especially in developed nations. The current configuration of animal agriculture is deemed unsustainable for the existing population, which will only become aggravated as the world's population and income continue to expand.

Additionally, animal agriculture also raises a number of issues related to public health. The escalation of animal farming has led to an increased occurrence of zoonotic illnesses, primarily attributed to genetically similar animals being housed and transported in high densities. This close proximity facilitates the mutation of pathogens, leading to an elevated risk of zoonotic pandemics in environments such as wet markets and animal farms. Furthermore, a number of foodborne bacteria found in animal products have developed antibiotic resistance as a result of the long-term usage of prophylactic antibiotics in animal husbandry. Data from China [22], for instance, indicate that 18% of market- and shop-bought meat samples had Salmonella contamination, with 88% of these samples exhibiting resistance to at least one antimicrobial and 58% displaying multi-drug resistance. Similarly, in Italy, 88% of Campylobacter identified in poultry samples were resistant to at least one tested antimicrobial. In South Africa, *E. coli* isolates from meat have likewise shown alarming rates of multi-drug resistance. Conventional meat production increases the likelihood of pandemic risk and contributes to the issue of antibiotic resistance.

Furthermore, the excessive consumption of animal products is associated with various personal health issues, including heart disease and cancer [23]. Notably, eating red and processed meat is mainly linked to these serious health issues. In 2015, the World Health Organization designated processed and red meat to be carcinogenic. Specifically, red meat is classified as Group 2A, which indicates positive associations and strong mechanistic evidence linking red meat consumption to colorectal cancer. Processed meat is classified as Group 1, suggesting resounding evidence from epidemiological studies that processed meat causes cancer.

Despite the pertinence of this information to public health, there is a lack of estimates regarding the proportion of consumed meat that is processed. However, a rough estimate can be constructed based on available data. In 2020, the worldwide meat sales volume was 224 million tons for processed meat, 30 million tones for beef and veal, and 66 million tons for pork, according to the Good Food Institute's study of Euromonitor 2021 [24] data obtained from trade sources and national statistics. While there may be some overlay between processed meat and the beef/pork categories, a tentative estimation suggests that processed and/or red meat accounts for about half of the world's meat sales. Consequently, a significant amount of the meat is known to have carcinogenic qualities. Hence, industrial animal agriculture presents challenges within the food system concerning the environment, animal welfare, and human health. Despite the longstanding acknowledgment of these concerns, meat consumption has exhibited a steady increase since the 1960s. Although part of this growth can be attributed to global population expansion during the same period, the ongoing rise in global demand is anticipated, driven by an increasing number of customers in developing nations incorporating more meat into their daily meals. Many of the associated issues, particularly those related to health, are predominantly linked to the overconsumption of meat, which is mainly prevalent in advanced nations.

The increasing worldwide consumption of animal products, despite the adverse outcome, underscores that consumers' dietary preferences are not primarily influenced by ethical or environmental considerations. Instead, the main determinants of food choices are factors such as price, taste, health benefits, and convenience.

A promising approach to addressing these concerns involves plant-based animal product alternatives (PB-APAs) substituting the consumption of animal products. PB-APAs are goods designed to replicate the attributes of animal products, including their appearance, flavor, aroma, functionality, and cooking experience. Oatley oat milk and Beyond Meat are two well-known PB-APA examples. It is essential to distinguish PB-APAs from whole plant foods, which encompass minimally processed plant foods such as fruits, vegetables, legumes, nuts, and seeds. Unlike whole plant foods, PB-APAs prioritize consumer preferences for flavor familiarity and convenience by attempting to replicate the sensory qualities of animal product. Adopting PB-APAs provides consumers to reduce meat consumption while maintaining the culinary experiences they are accustomed to and enjoy. PB-APAs offer significant advantages over animal products, contributing to both public health and environmental sustainability. However, critics argue that PB-APAs may be perceived as less sustainable and healthful than a plant-based diet that just consists of whole foods [25].

Critics of PB-APAs contend that while these products may be an improvement over animal products, they still fall short of a straightforward diet of entire plant foods with little to no processing. In response to this argument, two key points merit consideration. Firstly, even if it is acknowledged that whole plant foods offer superior health or environmental outcomes compared to PB-APAs, the disparity between these two options is significantly smaller than the disparity between both option and a conventional meat-based diet. The significant orders of magnitude disparity in environmental impacts makes this clear. For any analyses indicating that PB-APAs are at least 50% less environmentally destructive than animal products, it logically follows that the gap between PB-APAs and animal products is larger than the difference between PB-APAs and whole plant foods. The majority of these assessments really indicate that the efficiency difference between PB-APAs and animal products is more like 90% than 50%.

Secondly, PB-APAs are designed to serve as substitutes for animal products, specifically meat, fish, eggs, and dairy—not substitutes for fruits and vegetables. Their physical forms, sensory characteristics, and product labels closely resemble those of traditional animal-based products. Consequently, it is probable that individuals purchasing PB-APAs are incorporating them into their diets as replacements for animal products instead of being alternatives for whole plant foods. The scenario of serving both a plant-based burger and a meat burger in the same dish without additional sides [26] appears improbable. Instead, it is more plausible that someone opting for a plant-based burger is choosing it as a substitute for the meat burger while also adding other accompaniments such as beans or salad.

The primary environmental metric frequently employed in assessments was greenhouse gas emissions, specifically measured by global warming potential. Early comparisons, such as those between plant-based animal product alternatives (PBAPAs) and pork products, revealed a substantial difference in their contributions to global warming. A study [27] determined that the pork supply chain demonstrated a 6.4-fold greater impact on global warming than the PBAPA supply chain. More comprehensive life cycle assessments encompassing a broader range of product categories were subsequently conducted. Nijdam [28] conducted a review of life cycle analyses examining various protein sources, disclosing the carbon footprint associated with each. The results showed that meat alternatives have substantially smaller footprints than animal products. The carbon footprint of entirely plant-based meat alternatives ranged from 1 to 2 kg CO_2e per kg of product, whereas those that contained eggs showed a footprint of 3–6 kg. In comparison, poultry, pork, and beef displayed carbon footprints ranging from 2 to 6 kg, 4 to 11 kg, and 9 to 120 kg, respectively. Consequently, plant-based products demonstrated up to 120 times greater carbon efficiency than their animal-based counterparts.

These findings are supported by recent life cycle analysis. Saget [29] compared pea-protein balls with Brazilian or Irish beef in attributional life cycle evaluations. The study revealed that the processing of pea protein exhibited a smaller environmental impact across all 16 environmental categories assessed, including an 89% reduction in global warming potential. The authors argued that a mere 5% substitution of German beef consumption with pea proteins might result in an annual reduction of 8 million tons of CO_2e emissions, approximately 1% of Germany's total annual emissions. According to Saget [29], when compared to beef burgers, PBAPAs are linked to 82–87% less climate impact per nutrition density unit. Furthermore, the authors asserted that when taking into account the carbon opportunity cost of land, the climate change benefits of PBAPAs rise by additional 25–44%. In the UK, transitioning from beef burgers to vegetable patties is said to have the potential to save 9.5–11 million tons of CO_2e annually, constituting 2.4% of the UK's territorial emissions.

Many studies utilize product nutrition labels as the primary tool for evaluating the nutritional contents of plant-based animal product alternatives (PBAPAs) and their animal-based equivalents. Alessandri [30] analyzed 207 PBAPAs and 226 meat items from 14 UK retailers. The result revealed that PBAPAs exhibited much greater amounts of fiber and salt but significantly lower levels of protein, saturated fat, and energy density. Applying the UK's Nutrient Profiling Model, the study discovered that 40% of meat products were classified as "less healthy" compared to only 14% of PBAPAs. Moreover, compared to only 20% of PBAPAs, 46% of meat products were deemed high in total fat, saturated fat, or salt. The authors concluded that PBAPAs had better nutritional profiles than meat, while there is a need to lower the salt concentration.

Similarly, Petersen [31] used Ofcom's A-score, which assesses the presence of various "nutrients to limit," including saturated fat, sodium, sugar, and overall calories, to evaluate the healthiness of new meat and meat substitutes in the German market. The analysis revealed that PBAPAs were far healthier than animal products since they had much lower quantities of nutrients to limit.

In the United States, Harnack [32] conducted a study comparing the nutritional characteristics of 37 PBAPAs and beef in the ground beef category. The findings indicated that PBAPAs were low in saturated fats and rich sources of dietary fiber, iron, manganese, copper, folate, and niacin. In contrast to ground beef, the products were found to have a higher salt content and lower levels of protein, zinc, and vitamin B12. The results of another study that compared seven PBAPA burgers to beef burgers in the USA yielded similar conclusions [33]. While plant-based burgers exhibited similar macronutrient profiles to 80% lean beef burgers—often having similar or higher protein levels, lower fat and saturated fat levels, and increased fiber content—the authors highlighted higher salt levels and reduced Iron, calcium, and protein bioavailability in PBAPAs. These studies suggest that, despite differing nutritious profiles, unequivocally declaring the overall health superiority of either PBAPAs or meat is not straightforward.

Various studies employed computer modeling to project the nutritional consequences of substituting animal products with plant-based animal product alternatives (PBAPAs) in specific populations. Vatanparast [34] conducted a modeling exercise to evaluate the effects of increasing PBAPA consumption by 100% while concurrently reducing red and processed meat consumption by 50% in Canada. The simulation showed decreased protein, cholesterol, zinc, and vitamin B12 levels and increased intake of fiber, polyunsaturated fatty acids, magnesium, and dietary folate equivalents. Utilizing Nutrient Rich Food scores, the authors concluded that in comparison to baseline diets, the PBAPA-based diet's total nutritional value was favorable. Farsi [35] employed a modeling approach to examine the nutritional implications of replacing meat with PBAPAs in the UK. The results revealed that switching to PBAPAs led to a decrease in the intake of protein, lipid, saturated fat, iron, and vitamin B12 and an increase in the intake of carbohydrates, fiber, sugars, and sodium. The authors suggested that PBAPAs, which are strong in fiber, protein, and micronutrients and low in salt and sugar, can be a nutritious alternative to meat. Additionally, they recommended fortifying PBAPAs with iron and B12, alongside efforts to reduce sugar and salt content, considerations that manufacturers and policymakers should incorporate.

Finally, numerous studies simultaneous modelled environmental and nutritional outcomes associated with various protein sources. Mertens [36] observed that plant-based animal product alternatives (PBAPAs), particularly fortified products, might improve nutrient quality while reducing environmental effects after seeing that diets optimized for nutritional quality required a 50% drop in meat consumption. In examining the effects of substituting meat and dairy with PBAPAs in a cohort of young Dutch females, they discovered that this replacement led to a 30% reduction in saturated fat intake without compromising total iron intake. The authors acknowledged that, while the iron content of the plant-based diet was higher overall, it was inclined to be derived from less bioavailable sources.

Overall, the existing literature consistently affirms that plant-based animal product alternatives (PBAPAs) have lower levels of fat, saturated fat, cholesterol, and calories than animal products. However, there is a noted variance in protein, iron, and B12 content, with some PBAPA products displaying lower or less bioavailable levels. Despite concerns regarding elevated salt levels in certain products, PBAPAs generally showcase higher levels of dietary fiber and various micronutrients. Consequently, PBAPAs present numerous positive health attributes from a nutritional standpoint and offer the potential for further development to enhance overall nutritional profiles.

The literature also presents evidence supporting additional health benefits associated with plant-based animal product alternatives (PBAPAs), encompassing cholesterol reduction and positive impacts on gut health. Crimarco [37] demonstrated that PBAPA consumption, when compared to meat consumption, was linked to significantly lower concentrations of LDL cholesterol. Moreover, Ritchie [38] established that substituting PBAPA for animal products in current diets could avert as many as 52,700 premature deaths annually. The analysis indicated a higher level of PBAPA uptake resulted in a more significant avoidance of premature deaths, with diet-related health risks—such as cancer, stroke, and coronary heart disease—accounting for more than 85% of the prevented fatalities, rather than weight management. The authors advocate for a shift towards PBAPAs for public health. In summary, the evidence in this domain suggests that PBAPAs can contribute to cholesterol reduction, enhance gut health, and play a role in preventing premature deaths. When compared to animal products, PBAPAs consistently have better nutritional profiles, are quite effective in promoting muscle growth and weight loss, and can be customized to treat particular health issues. Additionally, they offer cholesterol-lowering benefits and promote gut health. Ongoing study endeavors aimed at enhancing the health attributes of PBAPAs involve identifying ingredients and processes to optimize protein and fiber content, improve vitamin content, and reduce antinutrient content. Future research should explore strategies to reduce salt intake while increasing protein, iron, and vitamin B12.

References

1. Human Verification. https://www.semanticscholar.org/paper/An-investigation-of-the-formulation-and-nutritional-Bohrer/d1d024bb9b5ea91578acb98abc9e34e71fd8a271 (n.d.).
2. Yang Y, Zheng Y, Ma W, Zhang Y, Sun C, Fang Y. Meat and plant-based meat Analogs: nutritional profile and in vitro digestion comparison. Food Hydrocoll. 2023;143:108886. https://doi.org/10.1016/j.foodhyd.2023.108886.
3. Xie Y, Cai L, Zhao D, Liu H, Xinglian X. Real meat and plant-based meat analogues have different in vitro protein digestibility properties. Food Chem. 2022;387:132917. https://doi.org/10.1016/j.foodchem.2022.132917.
4. Lantern | Papers: The Green Revolution. https://www.lantern.es/lanternpapers-eng/the-green-revolution-2020 (n.d.).
5. Singh M, Trivedi N, Enamala MK, Chandrasekhar K, Parikh P, Nikolova MP, Chavali M. Plant-based meat analogue (PBMA) as a sustainable food: a concise review. Eur Food Res Technol. 2021;247(10):2499–526. https://doi.org/10.1007/s00217-021-03810-1.
6. ProVeg España. Productos De Origen Vegetal En España | Informes Y Encuestas; 2022. https://proveg.com/es/que-hacemos/compromiso-corporativo/productos-de-origen-vegetal-en-espana/.
7. Curtain F, Grafenauer S. Plant-based meat substitutes in the flexitarian age: an audit of products on supermarket shelves. Nutrients. 2019;11(11):2603. https://doi.org/10.3390/nu11112603.
8. Fresán U, Mejía MA, Craig WJ, Jaceldo-Siegl K, Sabaté J. Meat Analogs from different protein sources: a comparison of their sustainability and nutritional content. Sustainability. 2019;11(12):3231. https://doi.org/10.3390/su11123231.
9. Gómez Ramirez BD, Echeverri Osorio MC, Rojas Pinilla AM. Evaluación del etiquetado de alimentos dirigidos a población vegetariana en Colombia. Revista Española de Nutrición Humana y Dietética. 2021;25(4):403–18.
10. Boukid F, Castellari M. Veggie burgers in the EU market: a nutritional challenge? Eur Food Res Technol. 2021;247(10):2445–53. https://doi.org/10.1007/s00217-021-03808-9.
11. Medawar E, Hühn S, Villringer A, Veronica Witte A. The effects of plant-based diets on the body and the brain: a systematic review. Transl Psychiatry. 2019;9(1) https://doi.org/10.1038/s41398-019-0552-0.

12. Felman A. What to know about cardiovascular disease; 2023. https://www.medicalnewstoday.com/articles/257484.
13. Cancer: Carcinogenicity of the Consumption of Red Meat and Processed Meat. https://www.who.int/news-room/q-a-detail/cancer-carcinogenicity-of-the-consumption-of-red-meat-and-processed-meat (n.d.).
14. American Heart Association. Eating more plant protein and dairy instead of red meat may improve heart health. https://newsroom.heart.org/news/eating-more-plant-protein-and-dairy-instead-of-red-meat-may-improve-heart-health (n.d.).
15. The American Journal of Clinical Nutrition | ScienceDirect.com by Elsevier. https://www.sciencedirect.com/journal/the-american-journal-of-clinical-nutrition (n.d.).
16. Sun Z, Scherer L, Tukker A, Spawn-Lee SA, Brückner M, Gibbs HK, Behrens P. Dietary change in high-income nations alone can Lead to substantial double climate dividend. Nat Food. 2022;3(1):29–37. https://doi.org/10.1038/s43016-021-00431-5.
17. Morach B, et al. Food for thought: the protein transformation. BCG Global; 2023. https://www.bcg.com/publications/2021/the-benefits-of-plant-based-meats.
18. UBS Editorial Team. Food-revolution. Insights; 2023. https://www.ubs.com/global/en/wealth-management/insights/chief-investment-office/sustainable-investing/2019/food-revolution.html#:~:text=Sweeping%20innovations%2C%20including%20trends%20in,How%20will%20supply%20chains%20adapt%3F.
19. Jähn S, Furchheim P, Strässner A-M. Plant-based meat alternatives: motivational adoption barriers and solutions. Sustainability. 2021;13(23):13271. https://doi.org/10.3390/su132313271.
20. Flint M, Bowles S, Lynn A, Paxman J. Novel plant-based meat alternatives: future opportunities and health considerations. Proc Nutr Soc. 2023;82(3):370–85. https://doi.org/10.1017/s0029665123000034.
21. McClements DJ. Ultraprocessed plant-based foods: designing the next generation of healthy and sustainable alternatives to animal-based foods. Compr Rev Food Sci Food Saf. 2023;22(5):3531–59. https://doi.org/10.1111/1541-4337.13204.
22. Xu Z, Wang M, Zhou C, Guimin G, JingZhen L, Hou X, Wang M, Wei P. Prevalence and antimicrobial resistance of retail-meat-borne salmonella in southern China during the years 2009–2016: the diversity of contamination and the resistance evolution of multidrug-resistant isolates. Int J Food Microbiol. 2020;333:108790. https://doi.org/10.1016/j.ijfoodmicro.2020.108790.
23. Al-Shaar L, Satija A, Wang DD, Rimm EB, Smith-Warner SA, Stampfer MJ, Hu FB, Willett WC. Red meat intake and risk of coronary heart disease among US men: prospective cohort study. BMJ. 2020:m4141. https://doi.org/10.1136/bmj.m4141.
24. Passport. Euromonitor. https://www.euromonitor.com/our-expertise/passport (n.d.).
25. Scipioni J. Impossible CEO on critics who say plant-based meat is unhealthy: 'It's Bull'. CNBC; 2021. https://www.cnbc.com/2020/07/21/impossible-ceo-on-criticism-that-plant-based-meat-is-unhealthy.html.
26. Impact of New Plant-Based Protein Alternatives on U.S. Beef Demand – Full Report | AgManager.info. https://www.agmanager.info/livestock-meat/meat-demand/meat-demand-research-studies/impact-new-plant-based-protein-0 (n.d.).
27. Zhu X, Van Ierland EC. Protein chains and environmental pressures: a comparison of pork and novel protein foods. Environ Sci. 2004;1(3):254–76. https://doi.org/10.1080/15693430412331291652.
28. Nijdam D, Rood T, Westhoek H. The price of protein: review of land use and carbon footprints from life cycle assessments of animal food products and their substitutes. Food Policy. 2012;37(6):760–70. https://doi.org/10.1016/j.foodpol.2012.08.002.
29. Saget S, Costa MP, Santos CS, Vasconcelos MW, Gibbons J, Styles D, Williams ML. Substitution of beef with pea protein reduces the environmental footprint of meat balls whilst supporting health and climate stabilisation goals. J Clean Prod. 2021;297:126447. https://doi.org/10.1016/j.jclepro.2021.126447.

30. Alessandrini R, Brown MK, Pombo-Rodrigues S, Bhageerutty S, He FJ, MacGregor GA. Nutritional quality of plant-based meat products available in the UK: a cross-sectional survey. Nutrients. 2021;13(12):4225. https://doi.org/10.3390/nu13124225.
31. Petersen T, Hartmann M, Hirsch S. Which meat (substitute) to buy? Is front of package information reliable to identify the healthier and more natural choice? Food Qual Prefer. 2021;94:104298. https://doi.org/10.1016/j.foodqual.2021.104298.
32. Harnack L, Mork S, Valluri S, Weber C, Schmitz K, Stevenson J, Pettit J. Nutrient composition of a selection of plant-based ground beef alternative products available in the United States. J Acad Nutr Diet. 2021;121(12):2401–8. https://doi.org/10.1016/j.jand.2021.05.002.
33. Cereals & Grains. https://www.cerealsgrains.org/publications/cfw/2020/July-August/Pages/CFW-65-4-0045.aspx (n.d.).
34. Vatanparast H, Islam N, Shafiee M, Dan Ramdath D. Increasing plant-based meat alternatives and decreasing red and processed meat in the diet differentially affect the diet quality and nutrient intakes of Canadians. Nutrients. 2020;12(7):2034. https://doi.org/10.3390/nu12072034.
35. Farsi DN, Uthumange D, Muñoz-Muñoz JL, Commane DM. The nutritional impact of replacing dietary meat with meat alternatives in the UK: a modelling analysis using nationally representative data. Br J Nutr. 2021;127(11):1731–41. https://doi.org/10.1017/s0007114521002750.
36. Mertens E, Biesbroek S, Dofková M, Mistura L, D'Addezio L, Turrini A, Dubuisson C, et al. Potential impact of meat replacers on nutrient quality and greenhouse gas emissions of diets in four European countries. Sustainability. 2020;12(17):6838. https://doi.org/10.3390/su12176838.
37. Crimarco A, Springfield S, Petlura C, Streaty T, Cunanan K, Lee J, Fielding-Singh P, et al. A randomized crossover trial on the effect of plant-based compared with animal-based meat on trimethylamine-N-oxide and cardiovascular disease risk factors in generally healthy adults: study with appetizing plantfood – meat eating alternative trial (SWAP-MEAT). Am J Clin Nutr. 2020;112(5):1188–99. https://doi.org/10.1093/ajcn/nqaa203.
38. Ritchie H, Reay D, Higgins P. Potential of meat substitutes for climate change mitigation and improved human health in high-income markets. Front Sustain Food Syst. 2018;2 https://doi.org/10.3389/fsufs.2018.00016.

Chapter 5
Scientific Advances in Meat Substitute Technologies

Recent advancements in food technology have revolutionized the production of meat substitutes, offering sustainable and ethical alternatives to traditional animal-based proteins. Scientists are exploring novel approaches, such as cellular agriculture, which involves cultivating animal cells in controlled laboratory environments to create cultured meat products [1]. Another promising avenue is the development of plant-based proteins that closely mimic the taste, texture, and nutritional profile of conventional meat [2]. Researchers are utilizing sophisticated techniques like extrusion, high-moisture extrusion, and 3D printing to create complex protein structures that replicate the fibrous nature of muscle tissue [3]. Additionally, fermentation processes are being employed to produce mycoprotein and other fungal-based meat alternatives [4]. These cutting-edge innovations address environmental concerns associated with livestock farming and cater to the growing consumer demand for more sustainable and cruelty-free food options. The development of meat substitutes represents a significant step towards a more sustainable and ethical food system, aligning with global efforts to reduce the environmental impact of food production and meet evolving consumer preferences.

A comprehensive life cycle assessment published in *Science* found that meat substitutes generally have a significantly lower environmental impact compared to conventional meat production [5]. The study showed that plant-based alternatives can reduce greenhouse gas emissions by up to 90% and land use by up to 99% compared to beef production. Research on consumer acceptance of meat substitutes has shown promising results. A study in Appetite found that providing information about the environmental and health benefits of cultured meat increased consumer acceptance and willingness to try these products [6]. Additionally, a market analysis by Boston Consulting Group (2021) projected that alternative proteins could account for 11% of the global protein market by 2035, indicating growing consumer interest and market potential [7]. These scientific advancements and studies demonstrate the significant progress and potential of meat substitutes in addressing

S. Ray, S. Suri, *The Sustainable Plate: Unveiling the Science of Meat Substitutes and Impact on Global Health*, https://doi.org/10.1007/978-3-031-94604-2_5

environmental concerns and meeting consumer demands for sustainable and ethical food options.

5.1 Latest Advancements in Plant-Based Food Processing

Recent advancements in plant-based food processing technologies have significantly increased the ability to create meat and seafood analogues that closely replicate the sensory attributes—texture, flavor, and appearance—of traditional animal-based foods. While extrusion technology has served as a foundational process in food manufacturing for decades, new variations, such as high-moisture extrusion, have further optimized this technique, allowing manufacturers to produce plant-based products with enhanced textural characteristics that more closely resemble the chewiness and tenderness found in conventional meat [8].

In addition to extrusion, several emerging technologies, including ohmic heating, shear cell processing, electrospinning, and 3D printing, are shaping the evolution of alternative proteins by providing novel approaches to enhance textural fidelity, nutritional quality, and functional properties [9]. Ohmic heating—also referred to as electrical resistance heating—employs alternating currents to heat food rapidly and uniformly, making it particularly effective for pasteurization and sterilization in food production [10]. When applied to plant-based proteins, ohmic heating minimizes nutrient degradation—a common risk associated with conventional thermal treatments—thus preserving key sensory qualities such as flavor and texture [11]. Research indicates that ohmic heating can also improve plant protein functionality, including emulsification and gelation, which are critical for achieving the desired mouthfeel in meat alternatives.

Another promising innovation, shear cell technology, a form of high-pressure processing (HPP), applies controlled pressure to create fibrous textures that resemble meat's structural complexity. When combined with precise deformation, shear cell processing enables the production of stable, plant-based products with consistent fibrous structures, maintaining their authenticity even after cooling [12]. Companies like Rival Foods are pioneering shear cell technology to produce whole-cut plant-based products with stable and appealing textures, effectively leveraging the method's capacity to modify plant protein structures at a microscopic level [13]. Electrospinning, a scaffold-based technology, restructures plant-based proteins into nanofibers through the application of an electric field. This process replicates the intricate muscle tissue textures found in animal products, creating complex structural integrity necessary for high-fidelity meat analogues. Although electrospinning is still in its developmental phase for large-scale food applications, it holds substantial promise for advancing the production of authentic plant-based meat textures [14]. Companies like Meatable, in partnership with VIVOLTA, are actively exploring the feasibility of electrospun scaffolds to help bridge the sensory gap between plant-based and traditional meat [15].

Further expanding the scope of customization, 3D printing offers a high degree of control by layering plant proteins to replicate the structure and texture of whole cuts of meat and seafood. This technology allows for precision in product design, enabling the creation of plant-based products with textures and flavors tailored to consumer preferences [16]. For example, Revo Foods has been applying 3D printing to produce seafood analogues, including plant-based tuna spreads and salmon, that simulate the sensory experience of conventional seafood [15].

Alongside these physical processing innovations, recent advancements in enhancing *protein functionality* have also explored both physical and chemical modification techniques [17]. Chemical modifications, such as glycosylation, phosphorylation, pH adjustments, and deamidation, are effective in altering protein properties, yet they face growing scrutiny due to consumer demand for "clean label" products. The use of certain chemical agents introduces challenges for scaling up production because of potential toxicity. Physical modification methods, including freeze-thawing, thermal treatment, and hot-melt extrusion, can effectively alter protein structure but often require substantial energy inputs, which limits their cost-effectiveness. To address these challenges, a growing interest in innovative non-thermal and non-chemical methods has emerged, including precision fermentation, high-pressure processing (HPP), 3D extrusion, pulsed electric field (PEF), and ultrasonic technology. Precision fermentation allows to produce recombinant proteins with enhanced functionality and has demonstrated effectiveness in reducing off-flavors in plant proteins by generating ester compounds through specific yeast co-cultures. Non-thermal methods like HPP and PEF further contribute to improving texture by inducing protein denaturation and gelation while maintaining sensory quality. Additionally, advanced techniques like 3D extrusion and ultrasonic technology enable refined structural modifications in protein processing, supporting environmentally friendly practices and promoting sustainability in plant protein production.

5.2 Fungi-Based Meat Alternatives: Technological Advancements

Fungi-based mycoprotein production has gained significant recognition as a sustainable, protein-rich food source, primarily due to its efficient use of agro-industrial byproducts, such as seaweed waste, soy residues, date waste, and pea processing byproducts [18]. This process employs various microbial fermentation methods to transform these substrates into nutrient-dense mycoprotein, with production techniques carefully chosen based on scale, substrate type, and the fungal strain used. Three main techniques dominate this field: submerged fermentation (SmF), solid-state fermentation (SSF), and the surface culture method (SCM), each offering unique advantages.

Submerged fermentation (SmF) is widely used in the large-scale production of mycoprotein and involves microbial growth in a liquid medium, which supports anaerobic or semi-anaerobic fermentation. Fungal strains like Fusarium venenatum are cultivated in a nutrient-rich medium filled with essential elements such as carbon, nitrogen, vitamins, and minerals. The fungal cells metabolize these nutrients within highly controlled bioreactors, where parameters like pH, temperature, oxygen, and agitation are precisely managed to optimize protein synthesis and biomass production. A notable advantage of SmF is the use of large bioreactors. Air-lift bioreactors, which rely on sterile air bubbles for mixing and aeration, enhance the delivery of oxygen to the fungal culture, facilitating extended hyphal growth that contributes to meat-like textures. These air-lift systems also operate with lower energy requirements compared to traditional stirred-tank bioreactors. After fermentation, the fungal biomass is isolated through centrifugation or filtration, then processed to improve flavor and texture, making it suitable for various food applications. SmF's consistency and efficient control make it ideal for large-scale commercial mycoprotein production, particularly for creating protein-rich substitutes for conventional meats.

Solid-state fermentation (SSF) offers an alternative approach, utilizing minimal liquid and fermenting the fungal culture on solid, insoluble substrates. This technique is particularly popular in the food industry due to its efficient nutrient utilization and lower water requirements compared to SmF. Common substrates for SSF include brewer's spent grain, grape waste, and other cereal grains, which serve as nutrient-dense environments that support the growth of protein-rich fungal biomass. In SSF, substrate preparation, fungal inoculation, and controlled incubation are key steps. Specific fungal strains, such as *Pleurotus albidus*, are introduced to the substrate, facilitating mycelial growth under regulated temperature, humidity, and aeration conditions. Extraction of mycoprotein involves dissolving the biomass in organic solvents like hexane, followed by filtration and centrifugation to purify the protein. SSF-produced mycoprotein has a high protein digestibility-corrected amino acid score (PDCAAS), often surpassing that of animal-based proteins like chicken or beef. However, while SSF is effective and environmentally efficient, it can be slower to scale for industrial production compared to SmF.

The surface culture method (SCM), though less scalable, is an effective technique for laboratory-scale production of mycoprotein. This technique involves growing fungal strains such as *Fusarium venenatum* on the surface of solid substrates like pea processing byproducts or date juice under controlled environmental conditions. The fungal biomass grows on the substrate's surface, allowing researchers to observe fungal behavior and nutrient needs more closely. SCM is particularly useful in experimental settings due to its simplicity and suitability for small-scale mycoprotein production. It provides researchers with a practical setup for testing different substrates and nutrient compositions to optimize protein yields. While SCM can achieve high protein content—up to 48% in dried fungal cells—it is resource-intensive and laborious, requiring careful post-incubation separation of mycoprotein via filtration and centrifugation. Consequently, SCM is not ideal for large-scale production but remains valuable for testing and development purposes.

Each of these fermentation methods offers distinct advantages for fungi-based mycoprotein production. Submerged fermentation (SmF) stands out for large-scale applications due to its precise control overgrowth conditions and high yield potential, while solid-state fermentation (SSF) is valued for its efficient use of resources and nutrient-rich output, particularly in high-fiber substrates. The surface culture method (SCM), while more limited in scalability, is invaluable for laboratory research and development, contributing to our understanding of fungal behavior and the refinement of fermentation conditions. Collectively, these methods underscore the potential of fungi-based mycoprotein as a sustainable alternative to conventional protein sources, meeting rising global demand for nutritious, environmentally friendly food options.

5.3 Advancement in Lab Grown Meat Production

The global population growth and rising meat demand pose significant environmental and ethical challenges, with traditional meat production contributing to high greenhouse gas emissions, extensive land use, and water consumption [5]. Cultured meat offers a sustainable alternative, producing animal protein without live animals, reducing GHG emissions by 96%, water use by 96%, and land use by 99% [19]. Cultured meat also improves efficiency with reduced nitrogen emissions and a smaller carbon footprint, surpassing even chicken in crop-to-meat conversion. As a transformative innovation, cultured meat addresses environmental and ethical concerns while emerging as a platform for delivering functional nutrients like nutraceuticals, prebiotics, and probiotics [20].

Cultured meat production involves a sophisticated methodology that combines cellular agriculture with tissue engineering [21]. The process primarily relies on skeletal muscle cells and adipocytes, with myosatellite cells being the most commonly utilized due to their regenerative capabilities and adaptability. Emerging alternative cell sources, such as embryonic stem cells and fibro-adipogenic progenitor cells, are gaining scientific interest, although myosatellite cells from adult animals remain the primary choice. The efficiency of cell isolation is influenced by several factors, including the age, sex, and tissue origin of the donor animal. For large-scale myosatellite cell propagation in cultured meat production, it is essential to replicate the natural physiological niche in vitro [22]. This requires cell growth media enriched with signaling molecules, extracellular matrix proteins like fibronectin and collagen, and controlled incubation systems to support cell proliferation, viability, and longevity. While serum-based media are commonly used for their proteins, minerals, and growth factors, serum-free formulations are actively being explored to reduce reliance on animal-derived components such as fetal bovine serum. Despite promising results, further research is needed to develop chemically defined media with comparable efficacy. Cultured meat production faces additional challenges, including optimizing large-scale manufacturing, replicating the structure and sensory qualities of conventional meat, and addressing high costs associated with growth factors and basal media [23]. Regulatory, socioethical, and consumer acceptance hurdles, as well as contamination risks from scaffolds, instruments, and cell culture infections, further complicate scalability. Advancing serum-free media, cost-effective production methods, and contamination control is essential for transforming cultured meat into a sustainable, economically viable alternative to traditional meat [24].

Advancements in technology are driving cultured meat towards large-scale commercialization by addressing critical production challenges. Precision fermentation emerges as a key innovation, enabling the development of serum-free media components such as human platelet lysate and insulin-like growth factor 1 (IGF-1), as well as scaffolds like collagen and alginate [25]. It also plays a role in enhancing sensory characteristics with heme and fatty acids and improving nutritional content through antioxidants and micronutrients. Progress in tissue engineering has introduced techniques such as hydrogel assembly, which facilitates the formation of contractile bovine muscle tissue with suitable physiological and sensory traits [26]. Additionally, cell sheet-based methods have produced biomimetic muscle tissues, replicating the flavor, texture, and nutritional profile of conventional meat. Brief magnetic field exposure has shown promise in promoting muscle cell proliferation, while microfluidic systems ensure product quality through precise bioanalysis. Furthermore, helical micromotors incorporating magnetic nanoparticles have advanced muscle tissue engineering, offering innovative solutions for scalable cultured meat production.

The advanced technologies such as cell culture, media development, scaffolds, bioreactors, and three-dimensional printing each play a critical role in creating sustainable and high-quality meat alternatives. For cell culture, robust cell lines capable of consistent performance are essential for large-scale production, with advancements in genetic engineering and microfluidic gene delivery systems improving cell line stability and proliferation. Serum-free media formulations are actively being developed to replace fetal bovine serum, reducing costs, and aligning with cultured meat's ethical goals. Innovative media formulations, such as recombinant albumin-enriched and chemically defined media, have demonstrated success in promoting cell growth and differentiation [27]. Scaffolds are crucial for providing structure to cultured meat, enabling cell attachment, growth, and differentiation. Biocompatible materials like decellularized plant tissues, synthetic polymers, and recombinant collagen are being explored. Bioreactors provide controlled environments for cell proliferation and differentiation, with technologies such as hollow fiber and single-use bioreactors supporting scalability and cost-efficiency [28]. Microcarriers and dynamic bioreactor systems further enhance scalability and quality. 3D printing has enabled the fabrication of structured cultured meat products like steak and sausages by combining bio-inks containing cells and biomaterials [29]. Recent breakthroughs include three-dimensional printed rib-eye steak and functional tissues mimicking native meat characteristics, showcasing the potential for widespread adoption of cultured meat technology. Multifaceted technological advancements driving the development of cultured meat are categorized into five core areas: 3D-printing, bioreactors, scaffolding, media, and cellular innovations

[30]. 3D-printing technologies have enabled breakthroughs such as the first 3D-printed rib-eye steak and customized meat textures using bioinks and hydrogels. Bioreactor innovations include single-use systems, co-culturing techniques, and novel reactor designs to enhance scalability and efficiency. Scaffolding has seen diverse material applications, ranging from plant-based and protein-based structures to edible and biodegradable supports. Media advancements focus on eliminating serum through innovations like Beefy-9 and closed-loop systems, supporting ethical and cost-effective cell growth. Cellular technologies are evolving through serum-free and genetically modified cell lines, microfluidic analysis, and repositories of farm animal cells. Collectively, these technological domains converge to make cultured meat a viable, scalable, and sustainable alternative to conventional animal agriculture.

Cultured meat holds immense potential to transform the food industry, especially with advancements in nutrition and functional food applications [31]. Future developments could see cultured meat enriched with bioactive compounds, vitamins, minerals, probiotics, and prebiotics, creating health-enhancing "functional cultured meat" products. Fortification with essential nutrients like vitamin B12, omega-3 fatty acids, branched-chain amino acids, and glutamine can boost cultured meat's nutritional profile, particularly for athletes and those with specific health needs. Additionally, plant-based fats may be added post-production to achieve a lipid profile like traditional meat. Innovative scaffolds, comprising polymers like collagen, provide structure while enhancing nutrient content. Though the field is in early stages, ongoing research on sustainability, regulatory issues, and consumer acceptance is crucial for successful commercialization. Overall, cultured meat's progress from 2013 to 2023 shows promises for a sustainable and ethical meat alternative, with potential applications ranging from nutrient-rich functional foods to a reliable protein source for space exploration.

5.4 Conclusion

Recent advancements in food science and technology have revolutionized meat substitute production, addressing environmental concerns, animal welfare, and the need for sustainable protein sources. Cellular agriculture, a key innovation, cultivates animal cells in controlled settings to produce cultured meat, eliminating the need to raise and slaughter animals. This approach integrates tissue engineering and biotechnology, utilizing scaffold-based cell growth, bioreactors, and serum-free media. Myosatellite cells, plant-based scaffolds, and precision fermentation enhance efficiency and scalability while reducing reliance on animal inputs. These developments offer sustainable, ethical alternatives to traditional meat, paving the way for a more humane and environmentally friendly food system.

Plant-based meat substitutes have advanced significantly, replicating the taste, texture, and nutrition of conventional meat. Technologies like high-moisture extrusion and 3D printing enable the creation of fibrous protein structures mimicking muscle tissue. Ingredients such as soy, wheat, and pea proteins are blended with oils and flavor enhancers for a meat-like experience. High-moisture extrusion aligns protein molecules into fibrous structures under controlled heating and shearing, while 3D printing offers precision in creating customized textures and shapes for diverse culinary applications. Plant-based meat substitutes have advanced significantly, replicating the taste, texture, and nutrition of conventional meat. Technologies like high-moisture extrusion and 3D printing enable the creation of fibrous protein structures mimicking muscle tissue. Ingredients such as soy, wheat, and pea proteins are blended with oils and flavor enhancers for a meat-like experience. High-moisture extrusion aligns protein molecules into fibrous structures under controlled heating and shearing, while 3D printing offers precision in creating customized textures and shapes for diverse culinary applications.

The adoption of meat substitute technologies presents significant environmental benefits, including reduced greenhouse gas emissions, lower water usage, and decreased land requirements compared to traditional livestock farming. Cultured meat, for example, has been shown to reduce land use by up to 95% and greenhouse gas emissions by up to 96% compared to conventional beef production. Similarly, plant-based and fungal-based meat substitutes contribute to resource conservation and provide a sustainable solution to meet global protein demands. Economically, the scalability of these technologies is improving. While the initial cost of cultured meat production remains high, advancements in bioreactor design, serum-free media, and automation are driving down costs, bringing these products closer to commercial viability. Plant-based proteins and fungal-based alternatives, already available on the market, continue to expand their reach as consumers embrace sustainable and health-conscious diets.

The future of meat substitute technologies lies in the integration of scientific disciplines, including synthetic biology, materials science, and food engineering. Continued research into improving taste, texture, and nutritional profiles will enhance consumer acceptance. Additionally, regulatory frameworks for novel foods will play a critical role in ensuring the safety and transparency of these products as they transition from the lab to the marketplace. By addressing key challenges such as scalability, cost, and consumer perception, these technologies have the potential to revolutionize the way we produce and consume protein, aligning with global efforts towards a more sustainable and equitable food system.

References

1. Benny A, Pandi K, Upadhyay R. Techniques, challenges and future prospects for cell-based meat. Food Sci Biotechnol. 2022;31(10):1225–42.
2. Su T, Le B, Zhang W, Bak KH, Soladoye PO, Zhao Z, et al. Technological challenges and future perspectives of plant-based meat analogues: from the viewpoint of proteins. Food Res Int. 2024;186:114351.

3. Dinali M, Liyanage R, Silva M, Newman L, Adhikari B, Wijesekara I, Chandrapala J. Fibrous structure in plant-based meat: high-moisture extrusion factors and sensory attributes in production and storage. Food Rev Intl. 2024:1–29.
4. Derbyshire E, Ayoob KT. Mycoprotein: nutritional and health properties. Nutr Today. 2019;54(1):7–15.
5. Poore J, Nemecek T. Reducing food's environmental impacts through producers and consumers. Science. 2018;360(6392):987–92.
6. Bryant C, Barnett J. Consumer acceptance of cultured meat: an updated review (2018–2020). Appl Sci. 2020;10(15):5201.
7. BostonConsultingGroup.Alternativeproteinmarkettoreach$290billionby2035;2021.Retrieved from https://www.bcg.com/press/23march2021-alternative-protein-market-reach-290-billion-by-2035.
8. Liu X, Yang C, Qin J, Li J, Li J, Chen J. Challenges, process technologies, and potential synthetic biology opportunities for plant-based meat production. LWT. 2023:115109.
9. Loura N, Gkartzou E, Trompeta AF, Konstantopoulos G, Klonos PA, Kyritsis A, Charitidis CA. Development of CNT-based nanocomposites with Ohmic heating capability towards self-healing applications in extrusion-based 3D printing technologies. C. 2023;9(4):111.
10. Varghese KS, Pandey MC, Radhakrishna K, Bawa AS. Technology, applications and modelling of ohmic heating: a review. J Food Sci Technol. 2014;51:2304–17.
11. dos Santos IF, Pimentel TC, da Cruz AG, Stringheta PC, Martins E, Campelo PH. Ohmic heating in food processing: an overview of plant-based protein modification. PRO. 2024;12(9):1800.
12. Bolumar T, Orlien V, Sikes A, Aganovic K, Bak KH, Guyon C, et al. High-pressure processing of meat: molecular impacts and industrial applications. Compr Rev Food Sci Food Saf. 2021;20(1):332–68.
13. Rival Foods. Rival Foods opens new production area to meet growing demand; 2024. Retrieved from https://rival-foods.com/2024/04/09/newproductionarea/.
14. Chakravorty P, Das AB. Advances in high moisture extrusion cooking for meat analogue with 3D and 4D printing approaches: a comprehensive review. Discov Food. 2024;4(1):137.
15. Bright Green Partners. Technological innovations in plant-based food processing; 2023. Retrieved from https://brightgreenpartners.com/technological-innovations-plant-based-food-processing/.
16. Dong H, Wang P, Yang Z, Xu X. 3D printing based on meat materials: challenges and opportunities. Curr Res Food Sci. 2023;6:100423.
17. Alasi SO, Sanusi MS, Sunmonu MO, Odewole MM, Adepoju AL. Exploring recent developments in novel technologies and AI integration for plant-based protein functionality: a review. J Agric Food Res. 2024:101036.
18. Majumder R, Miatur S, Saha A, Hossain S. Mycoprotein: production and nutritional aspects: A review. Sustain Food Technol. 2024;2(1):81–91.
19. Hubalek S, Post MJ, Moutsatsou P. Towards resource-efficient and cost-efficient cultured meat. Curr Opin Food Sci. 2022;47:100885.
20. Roy NK, Panda S, Dey G. Engineering a sustainable protein revolution: recent advances in cultured meat production. Food Bioeng. 2023;2(4):301–16.
21. Stephens N, Di Silvio L, Dunsford I, Ellis M, Glencross A, Sexton A. Bringing cultured meat to market: technical, socio-political, and regulatory challenges in cellular agriculture. Trends Food Sci Technol. 2018;78:155–66.
22. Shaikh S, Lee E, Ahmad K, Ahmad SS, Chun H, Lim J, et al. Cell types used for cultured meat production and the importance of myokines. Foods. 2021;10(10):2318.
23. Kirsch M, Morales-Dalmau J, Lavrentieva A. Cultivated meat manufacturing: technology, trends, and challenges. Eng Life Sci. 2023;23(12):e2300227.
24. Chu S, Haraguchi Y, Asahi T, Kato Y, Kondo A, Hasunuma T, Shimizu T. A serum-free culture medium production system by co-culture combining growth factor-secreting cells and l-lactate-assimilating cyanobacteria for sustainable cultured meat production. Sci Rep. 2024;14(1):19578.

25. Knychala MM, Boing LA, Ienczak JL, Trichez D, Stambuk BU. Precision fermentation as an alternative to animal protein, a review. Fermentation. 2024;10(6):315.
26. Namjoo AR, Abrbekoh FN, Saghati S, Amini H, Saadatlou MAE, Rahbarghazi R. Tissue engineering modalities in skeletal muscles: focus on angiogenesis and immunomodulation properties. Stem Cell Res Ther. 2023;14(1):90.
27. Lee DY, Lee SY, Yun SH, Jeong JW, Kim JH, Kim HW. Review of the current research on fetal bovine serum and the development of cultured meat. Food Sci Anim Resour. 2022;42:775–99.
28. Singh A, Kumar V, Singh SK, Gupta J, Kumar M, Sarma DK, Verma V. Recent advances in bioengineered scaffold for in vitro meat production. Cell Tissue Res. 2023;391(2):235–47.
29. Kulus M, Jankowski M, Kranc W, Golkar Narenji A, Farzaneh M, Dzięgiel P, et al. Bioreactors, scaffolds and microcarriers and in vitro meat production – current obstacles and potential solutions. Front Nutr. 2023;10:1225233.
30. Kulus M, Jankowski M, Kranc W, Golkar Narenji A, Farzaneh M, Dzięgiel P, Zabel M, Antosik P, Bukowska D, Mozdziak P, Kempisty B. Bioreactors, scaffolds and microcarriers and in vitro meat production – current obstacles and potential solutions. Front Nutr. 2023;10:1225233.
31. Kumar P, Sharma N, Sharma S, Mehta N, Verma AK, Chemmalar S, Sazili AQ. In-vitro meat: a promising solution for sustainability of meat sector. J Anim Sci Technol. 2021;63(4):693.

Chapter 6
Culinary Applications and Consumer Acceptance

The global culinary landscape has undergone a substantial transformation in recent years, characterized by an exponentially growing emphasis on meat alternatives. As society contends with the environmental, ethical, and health issues linked to conventional meat consumption, meat alternatives have transitioned from the periphery to a prominent position within the food market. This surge in demand for meat alternatives can be attributed to several converging factors. Increased awareness of the environmental impact of traditional meat production, alongside health concerns related to red meat consumption and the ethical issues of factory farming, has collectively spurred a significant shift in consumer behavior [1].

As individuals become more cognizant of the broader implications of their dietary choices, meat alternatives have gained prominence as a sustainable, ethical, and frequently healthier option, thereby enhancing their importance in the contemporary market.

6.1 Journey from Meat to Cultured Meats in Human Diet: A Brief History

The human predilection for meat dates back millions of years, playing a pivotal role in our evolutionary journey. This dietary shift was not merely a culinary innovation; it was a catalyst for significant physiological developments, including increases in body and brain size. Archaeological evidence, such as a 400,000-year-old hunting spear from Germany and indications of fire use nearly a million years ago, highlights the long-standing relationship between humans and meat [2]. As humanity evolved, so did our methods of acquiring meat, transitioning from opportunistic scavenging to organized hunting. Notable archaeological discoveries, such as the wooden spears from the Schöningen site in Germany, dating back approximately to

S. Ray, S. Suri, *The Sustainable Plate: Unveiling the Science of Meat Substitutes and Impact on Global Health*, https://doi.org/10.1007/978-3-031-94604-2_6

500,000 years, illustrate early examples of planned hunting expeditions [3]. Furthermore, evidence from a 2012 study indicates that Neanderthals were adept hunters, conducting large-scale deer hunts around 120,000 years ago [4].

The domestication of animals around 10,000 years ago marked another significant milestone, beginning in the Near East with goats and sheep and later spreading to other regions [5]. Livestock farming revolutionized human societies by providing a reliable food source and secondary products like milk, wool, and labor. This transformation facilitated population growth, settled lifestyles, and the rise of civilizations. Historically, meat has been a crucial component of human nutrition, providing essential nutrients such as protein, iron, and B vitamins. Protein is vital for tissue maintenance and repair, iron is crucial for hemoglobin production, and B vitamins support metabolism and brain function. Increased meat consumption has been linked to brain development, with evidence suggesting that a meat-rich diet contributed to the significant brain evolution of *Homo erectus*.

Global meat production and consumption continue to escalate, driven by factors such as population growth, economic development, and urbanization. According to a 2012 projection by the Food and Agriculture Organization (FAO) of the United Nations, global meat demand is expected to reach 455 million metric tons by 2050, representing a 76% increase from 2005 levels [6]. Similarly, global fish demand is anticipated to reach 140 million metric tons by 2050 [7, 8]. The surge in demand in meat consumption is predominantly seen in middle-income countries, such as China, while higher-income nations like the United Kingdom experience relatively stagnant or slightly decreasing meat consumption rates; conversely, in lower-income countries like India, meat consumption remains fairly constant [9]. This trend supports the hypothesis that the relationship between meat consumption and income follows an "inverted U-shaped" pattern: consumption rises with income up to a certain point, after which it plateaus or decreases [10]. This phenomenon may be explained by the increased awareness and concern for the environmental and ethical impacts of animal agriculture that often accompanies higher income levels.

Parallel to the increasing meat consumption narrative, vegetarianism has long-standing historical roots, often tied to religious, philosophical, or ethical beliefs. Ancient Indian and Mediterranean civilizations promoted vegetarian diets based on spiritual and health principles [11]. The term "vegetarian" emerged in the nineteenth century, coinciding with the formation of organized vegetarian societies in Europe. The twentieth century saw the rise of veganism, where individuals eschewed all animal-derived products, motivated by evolving concerns about animal welfare, health, and environmental sustainability. Simultaneously, the public health and environmental concern of meat consumption increased the demand for alternative options.

The twentieth century saw a notable rise in vegetarianism and veganism in Western societies, driving an increasing demand for meat substitutes [11]. Early iterations of these substitutes were often critiqued for their inadequate flavor, texture, and nutritional value when compared to traditional meat. However, these initial products laid the groundwork for subsequent innovations, setting the stage for the advanced meat alternatives available in today's market. Modern approaches to producing meat substitutes include plant-based meat (PBM) and cell-based meat (CBM), both of which aim to create food from non-animal sources [12]. Traditional PBMs, such as tofu, have a long history, but recent advancements have led to the commercialization of novel PBM products with improved sensory attributes [13]. Concurrently, the field of cellular agriculture has emerged, focusing on the production of commodities from cells rather than whole organisms [12]. This includes CBM, which involves cultivating muscle or fat cells to produce meat without the need for raising and slaughtering animals like cows, pigs, or chickens.

6.2 Inclusion and Versatility of Meat Substitutes in Various Cuisines and Recipes

Plant-based meat (PBM) products emulate the flavor, texture, and nutritional attributes of traditional meat while being derived from non-animal sources [14]. These PBM products can be broadly categorized into two flexible groups based on their historical development and technical sophistication: traditional and novel alternatives. Traditional meat analogs, which have been produced for thousands of years in Asia, include straightforward derivatives such as tofu and tempeh from soybeans and seitan from wheat.

In contrast, novel PBMs are engineered and marketed to closely replicate animal-based meat (ABM) in terms of taste, texture, and nutritional content [15]. Some PBM products may fall between these two categories, partially meeting the criteria for taste, texture, and nutritional equivalence to ABM. Novel PBM production involves three primary steps [16]. The first step is protein isolation and functionalization, where the target plant proteins are extracted and sometimes subjected to hydrolysis to enhance functionalities like solubility and cross-linking capacity. The second step is formulation, where the isolated plant proteins are combined with other ingredients to develop a meat-like texture. This mixture often includes food adhesives, plant-based fats, and flour. Additional nutrients are incorporated to match or exceed the nutrient profile of traditional meat. Finally, the formulated mixture undergoes various protein reshaping processes, such as stretching, kneading, trimming, pressing, folding, and extrusion, to achieve a meat-like texture.

Innovative technologies, such as shear cell technology, mycelium cultivation, 3D printing, and recombinant protein additives, are being employed to enhance the organoleptic properties of PBMs [14]. This technological advancement aims to create products that are nearly indistinguishable from traditional meat in terms of sensory experience.

Tofu, a soybean-based product, has its origins traced back to ancient China during the Han dynasty (206 BC–220 CE) [17]. Historical evidence, including depictions found in a Han dynasty tomb, attests to the early production methods of tofu. Its use as a meat substitute is documented in writings by Tao Gu (903–970), where tofu was colloquially referred to as small mutton, underscoring its value as an imitation of meat in Chinese cuisine [18]. Tofu gained widespread popularity during the Tang dynasty (618–907) and likely spread to Japan either in the later Tang period or early Song dynasty [17]. Tofu, also known as bean curd or soybean curd, is a creamy and nutritious soy product typically sold in block form. It boasts high protein content, has low fat, and is rich in calcium and iron. Made from soybeans, water, and a coagulant, tofu absorbs flavors well from spices, sauces, and marinades, making it a versatile ingredient in both Asian and Western vegetarian and vegan cuisines. It's an economical source of plant-based protein, often costing less than $2 for a block serving two to four people. While making tofu from scratch is possible but time-consuming, achieving the desired firm texture involves draining and pressing firm or extra-firm tofu before cooking. Freezing and thawing tofu can enhance its texture and flavor absorption, creating a chewier consistency akin to meat. Tofu can be prepared in various ways, including searing, grilling, frying, steaming, stewing, braising, baking, roasting, and even as a vegan alternative to scrambled eggs. Its neutral taste makes it ideal for integrating into dishes where it can take on the essence of savory sauces, aromatic spices, and various cooking methods. In stir-fries, tofu blends seamlessly with vegetables and sauces, creating dishes like tofu teriyaki or tofu and vegetable stir-fry. In curries, it adds a creamy texture while soaking up the rich flavors of coconut milk or tomato-based sauces, such as tofu tikka masala or tofu curry with vegetables. Tofu's adaptability extends to burgers and sandwiches, where it can be grilled, marinated, and served as patties or crispy strips in creations like tofu burgers or vegan banh mi. It also enhances soups and stews by absorbing broth flavors, and it complements salads with its protein-rich cubes or crumbles. Tofu finds a place in noodle dishes, stir-fried alongside vegetables and noodles in dishes like tofu pad Thai or tofu lo mein. For breakfast, tofu scrambles replace eggs in hearty dishes, offering a vegan alternative that can be enjoyed on its own or wrapped in breakfast burritos. Whether seared, grilled, baked, or scrambled, tofu's versatility not only enriches these recipes with plant-based protein but also delivers a satisfying and flavorful substitute for meat, catering to vegetarian, vegan, and health-conscious diets worldwide. Figure 6.1 illustrates bulk tofu and three tofu-based popular dishes from global cuisine.

Tempeh, a traditional Indonesian delicacy, is crafted from fermented soybeans through a natural culturing process that forms them into a solid cake [19]. Central to its production is the use of *Rhizopus oligosporus* or *Rhizopus oryzae* fungi, known as tempeh starter, which initiates the fermentation. Predominantly enjoyed on Java Island, tempeh stands out from tofu due to its whole-soybean composition, offering distinct nutritional benefits and a unique texture. During fermentation, soybeans are inoculated with *Rhizopus* spp. molds and spread thinly, left to ferment for about 24 to 36 hours at approximately 30 °C (86 °F). This process allows for spore germination and mycelium growth, binding the beans into a cohesive mass. The result is a firm cake with a white mycelium mat, characteristic of good quality tempeh. Its flavor, initially earthy and nutty, intensifies with aging. Tempeh's quality is assessed based on its mycelium density, firmness, and aroma, with good tempeh exhibiting a dense, uniform mycelium, barely visible beans, and a pleasant, slightly sweet aroma akin to mushrooms. Over-fermentation can darken its color and alter

Fig. 6.1 Raw tofu and tofu-based recipes. (Raw tofu: Attributed to Peter Griffin, License: CC0 Public Domain; Tofu stir fry: Attributed to Free Pik. The asset downloaded from Freepik; Tofu Buddha bowl: Attributed to Free Pik. The asset downloaded from Freepik; Tofu tikka masala: Attributed to Free Pik. The asset downloaded from Freepik)

its texture, distinguishing it from less desirable forms that crumble or lack proper mycelium formation. This meticulous process ensures tempeh maintains its prized qualities—nutritious, flavorful, and versatile—suitable for various culinary applications worldwide.

In culinary applications, tempeh is commonly prepared by slicing it into pieces, soaking it in brine or a savory sauce, and then frying it. On the island of Java, a traditional method involves marinating tempeh pieces in a mixture of ground garlic, coriander, turmeric, salt, and water before deep-frying them. This dish is often served with sambal ulek chili paste. Cooked tempeh can be enjoyed on its own or incorporated into a variety of dishes such as chili, stir-fries, soups, salads, sandwiches, and stews. Its complex flavor profile is often described as nutty, meaty, and reminiscent of mushrooms. Tempeh is also known for its excellent freezing capabilities and is widely available in Western supermarkets, ethnic markets, and health food stores. It can be steamed, marinated, thinly sliced, blackened, or crumbled into sauces and stews. Grated tempeh makes an excellent substitute for ground beef in dishes like tacos. When thinly sliced and deep-fried, it develops a crispy golden crust while maintaining a soft interior, making it ideal for marinating due to its sponge-like texture. Dried tempeh, whether cooked or raw, is more portable and less perishable, making it a convenient base for stews. Occasionally, diced tempeh will develop a white feathery fluff, which is the *Rhizopus* mold continuing to grow. This is normal and perfectly safe to eat.

Seitan, also known as wheat gluten or "wheat meat," is a high-protein meat substitute derived from gluten, the main protein in wheat [20]. Its origins can be traced back to ancient China, where it was developed by Buddhist monks as a meat substitute. Seitan has been a staple in East Asian vegetarian cuisine for centuries, particularly within Buddhist vegetarian traditions. Seitan has become increasingly popular in Western cuisine as a versatile plant-based protein option, serving as an alternative to other meat substitutes like tofu, edamame, or tempeh. It finds its way into a variety of dishes, including curries, stir-fries, and salads and is commonly used to create plant-based versions of traditional meat products such as ground "beef," deli slices, and sausages.

According to the USDA, 2 ounces (57 g) of crumbled seitan contains 100 calories, 17 g of protein, 0.5 g of fat, 6 g of carbohydrates, 1 g of fiber, 1.5 mg of iron (8% daily value or DV), and 440 mg of sodium (19% DV) [21]. Seitan is notable for being low in calories, saturated fat, and carbohydrates, while also serving as a good source of iron. In terms of protein, seitan is comparable to animal-based sources, with a 2-ounce serving containing as much protein as an equivalent amount of chicken breast. Compared to other plant-based proteins, seitan offers 4 g more protein per 2-ounce serving than 1 cup of edamame, 7 g more than 2 ounces of firm tofu, and 6 g more than 2 ounces of tempeh. However, seitan is an incomplete protein, lacking all essential amino acids, particularly lysine, which is abundant in animal products. To ensure a complete amino acid profile, it's beneficial to consume a variety of plant-based proteins throughout the day. Research, including a 2018 study published in Sustainable Protein Sources, suggests that seitan may be easier to digest compared to some other protein sources. However, it is unsuitable for those with celiac disease or gluten sensitivity due to its high gluten content. Store-bought seitan products may contain added salt and sugar, so it's important to check the nutrition facts and ingredients list. For those concerned about these additives, making seitan at home is a viable option. Homemade seitan can also be made with alternative flours such as beans, lentils, and chickpeas, which can alter its caloric content. Traditional seitan preparation involves combining wheat flour and water to create a dough, which is then kneaded and rinsed multiple times to remove starch, leaving the gluten. Nowadays, many people use vital wheat gluten to streamline the homemade seitan-making process. The dough can be shaped into various forms like loaves, noodles, or nuggets, and is typically poached in water or broth, often with seasonings like nutritional yeast or soy sauce for added flavor.

Historically, seitan was made by mixing wheat flour with water and kneading the mixture into a dough. This dough was then rinsed to remove the starch, leaving behind a mass of gluten [20]. The gluten would be cooked, often poached in a seasoned broth, before being used in various recipes. Today, the process has been simplified with the availability of vital wheat gluten, or gluten flour, which eliminates the need for the starch-removal step. Seitan can also be conveniently purchased pre-made in the refrigerated section of many supermarkets, typically near the tofu. Seitan itself has a neutral flavor, which some compare to unseasoned chicken breast or portobello mushrooms. This mild taste allows it to easily absorb the flavors of herbs, spices, sauces, and marinades, enhancing its umami and savory qualities. This adaptability makes it a favorite among those looking for a plant-based protein that can mimic the texture of meat, something tofu and tempeh do not typically provide.

The plant-based meat category remained small and relatively stagnant for many years, primarily catering to vegans and vegetarians. However, the market has expanded significantly in recent years, driven by the development of plant-based burgers and other products that closely mimic the taste and texture of conventional meat. This "biomimicry" approach started gaining momentum in 2012 with the launch of Beyond Meat's chicken strips [22]. The trend really accelerated with the 2016 introduction of the Impossible Burger and the Beyond Burger, both of which have achieved widespread acceptance and success, even being featured in mainstream fast-food outlets. These innovations have propelled the plant-based meat market into the mainstream, attracting not just vegetarians and vegans but also flexitarians and meat-eaters looking for healthier or more sustainable options.

As of 2020, over 800 companies were engaged in producing plant-based foods or had dedicated lines for such products [22]. By November 2022, more than 60 plant-based meat brands had surpassed $500,000 in sales. Notable companies include Beyond Meat, Impossible Foods, Eat JUST, Kellogg, Maple Leaf Foods, Pinnacle Foods, Quorn Foods, Tofurky, Gardein, MorningStar Farms, The Vegetarian Butcher, Sunfed, Gold & Green, Boca Foods, and Vbites Group. Beyond Meat has made a significant impact since its debut in 2012, offering products like burger patties, ground beef, sausage, and a relaunched chicken line, available in both grocery stores and restaurants. Impossible Foods is another major player, known for its Impossible Burger, sausage, and ground beef, available at chains like Burger King and retailers. Eat JUST, famous for its versatile vegan egg, is also venturing into cultivated meat, with plans to bring its cultivated chicken to the USA after regulatory approval in Singapore. Kellogg, known for its breakfast cereals, also produces plant-based meats under Incogmeato and MorningStar Farms. These brands offer a variety of products, from sausages to chicken and ground beef. Maple Leaf Foods owns Lightlife and Field Roast, which provide vegan deli meats, hot dogs, burgers, bacon, and holiday roasts. Pinnacle Foods, now owned by Conagra, includes Gardein, which offers plant-based chicken, fish filets, and prepared foods. Quorn Foods, owned by Monde Nissin Corporation, offers mycoprotein-based products, though some contain egg whites. Tofurky, a pioneer in plant-based meats since 1980, offers holiday roasts, deli slices, burgers, sausages, and tempeh. Gardein, also popular, provides fishless filets, chicken nuggets, and seasoned meats. MorningStar Farms has been a staple since 1975, with many vegan products, though some still contain egg whites. The Vegetarian Butcher, acquired by Unilever, offers vegan chicken, ground meat, and patties. Sunfed, based in New Zealand, produces chicken-free, beef-free, and bacon-free products. Gold & Green, from Finland, offers oat-based meats and has partnered with Taco Bell in the UK. Boca Foods offers mostly vegan products, though some contain animal ingredients, including turkey burgers and falafel bites. Vbites Group, based in the UK, has been producing a wide range of vegan meats and cheeses since 1993, including burger patties, chicken nuggets, and more unique items like plant-based patties and salmon slices. These companies have played a crucial role in making plant-based meat alternatives more mainstream, catering to both vegetarians and those seeking to reduce their meat consumption.

Mycoproteins derived from fungi have gained popularity due to their nutritious profile, cost-effective production, environmental advantages, and resilience to environmental challenges such as floods and droughts. The strain used to cultivate mycoprotein (*Fusarium venenatum* ATCC 2684) was discovered in the 1960s [23]. After extensive testing, mycoprotein received approval as a food protein source from the Ministry of Agriculture, Fisheries, and Food in the UK in 1984 and is now permitted for sale across all EU member states. Subsequent regulatory approvals followed in Switzerland, Norway, the USA, Australia, Japan, Thailand, Malaysia, and Canada. Marketed primarily in vegan and vegetarian food products under the brand name QuornTM, mycoprotein is produced on a large scale through fermentation, resulting in high-quality protein with minimal environmental impact. Figure 6.2 illustrates the timeline of commercialization of fungal protein [24].

Despite being often labeled as plant-based, mycoprotein comes from a different biological kingdom distinct from both plants and animals. Unlike mushroom-based products, mycoprotein is derived from the mycelium, the vegetative part of the fungus, similar to how roots relate to a tree. The search for sustainable alternative proteins during the "Green Protein Revolution" led British scientists to discover *Fusarium venenatum*, a micro fungus adept at converting starch into a fibrous, protein-rich substance [23]. This discovery paved the way for Marlow Foods to launch the first mycoprotein products under the Quorn brand in 1985 [22]. As Quorn's patents expired and the demand for alternative proteins grew due to

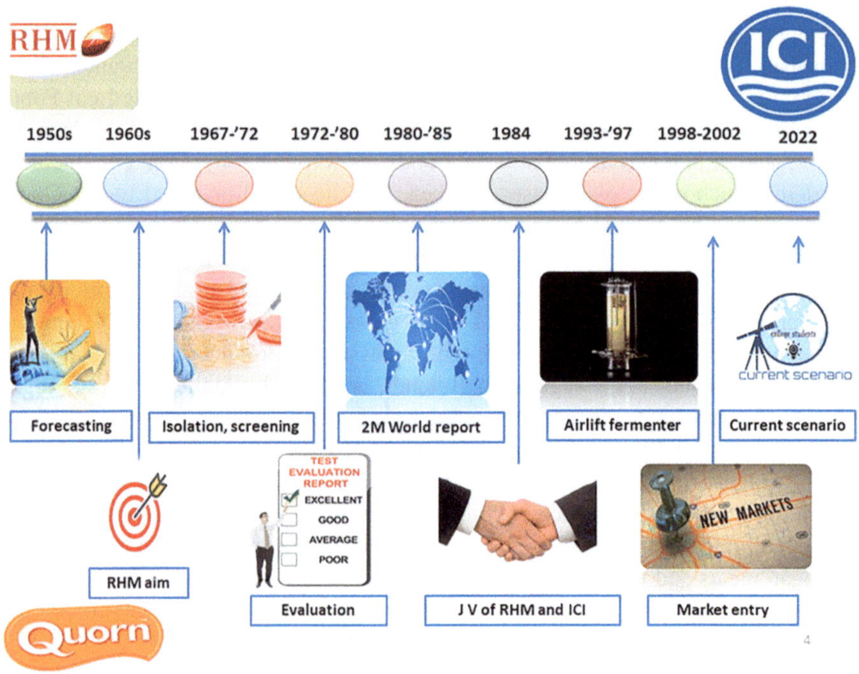

Fig. 6.2 Historical perspective of the development and commercialization of mycoprotein

population increases and environmental concerns, various startups began exploring mycoprotein production using innovative methods and new fungal strains. Swedish companies Mycorena and Millow, along with Scotland's ENOUGH, are notable for developing mycoprotein ingredients like Promyc™ and ABUNDA using submerged fermentation, while Millow uses dry fermentation techniques. These companies primarily operate in the business-to-business market. Meanwhile, companies like Spanish Libre and American MyForest Foods are directly targeting consumers with their mycoprotein-based retail products.

Mycoprotein, valued for its high protein content, increased fiber, and low levels of saturated fats, is a beneficial addition to a healthy diet [24]. Research indicates that mycoprotein can enhance satiety, help regulate blood sugar and cholesterol levels and offer several other nutritional benefits. It boasts a complete amino acid profile, ensuring it provides all essential amino acids necessary for dietary needs. The protein quality of mycoprotein is high, with a bioavailability-assessed amino acid score of 0.996, demonstrating its excellent production quality using ileostomy techniques. The fiber content is predominantly insoluble (88%) with 12% soluble fiber, consisting of a chitin and glucan matrix that is beneficial for intestinal health [24]. Glucans from grains and yeast in the fiber are known to aid fat metabolism and boost immune function. Mycoprotein is low in digestible carbohydrates, contributing to its composition of approximately one-third fat, and contains less than 1.5 g of both long- and short-chain saturated fatty acids per 100 g. It is rich in mono- and polyunsaturated fatty acids. Additionally, mycoprotein is a good source of several water-soluble B vitamins, including pyridoxine (0.1 mg), folate (114 μg), and cobalamin (0.72 μg). It is also rich in minerals such as zinc, phosphorus, calcium, iron, and potassium, making it a nutrient-dense food. According to the European Commission's nutrient claims, mycoprotein is recognized for its substantial mineral content and overall nutritional value.

Quorn, one of the pioneers in mycoprotein production, offers an extensive lineup of products including meat-free mince, burgers, sausages, and even ready-to-eat meals [25]. These products are designed to closely mimic the taste and texture of meat, making them an excellent substitute in traditional recipes. For instance, Quorn mince can be used in place of ground beef in lasagna, tacos, and Bolognese sauce, providing a healthy and environmentally friendly alternative. Other companies, such as Mycorena and ENOUGH, are focusing on the business-to-business supply chain, producing mycoprotein ingredients that food manufacturers can incorporate into their products. Mycorena's Promyc™ and ENOUGH's ABUNDA are produced using advanced fermentation techniques, ensuring high-quality protein content suitable for a variety of food applications. These ingredients can be used to create plant-based burgers, sausages, and other meat analogs that are both nutritious and sustainable [25]. Moreover, companies like MyForest Foods and Libre are targeting the retail market with innovative mycoprotein-based products [26]. MyForest Foods, for instance, offers mycelium bacon, which has been praised for its authentic taste and texture. Libre provides a range of ready-to-cook meals, such as stir-fries and casseroles, that utilize mycoprotein as the main protein source [27]. These products are not only convenient but also align with the growing consumer demand for plant-based alternatives. In the culinary setup, mycoprotein products are incredibly versatile. Chefs and home cooks alike are incorporating them into a myriad of dishes, from classic comfort foods like shepherd's pie and meatloaf to international cuisines such as curries and stir-fries. Their ability to absorb flavors makes them an excellent component in marinades and sauces. Additionally, mycoprotein's fibrous texture and high protein content make it a satisfying and healthy option for those looking to reduce meat consumption without compromising on taste or nutrition.

Historical records provide ample evidence of entomophagy, or the practice of eating insects, across various cultures and time periods [28]. Aristotle's "Historia Animalium" notes that cicadas were considered a delicacy in ancient Greece. Similarly, Pliny the Elder documented that the Romans enjoyed "cossus," a dish made from the larvae of longhorn beetles. Biblical references also mention insect consumption; the Old Testament allows the Hebrews to eat certain types of locusts (Leviticus 11:21–22), and the New Testament describes John the Baptist subsisting on locusts and wild honey (Mark 1:6) [29]. In pre-Hispanic Mesoamerica, entomophagy was a common practice [28]. Historical documents from the New Spain period indicate that the Aztecs consumed various insects, including winged ants, grasshoppers, mosquito eggs, and maguey worms [30]. Many of these insects remain popular in modern Mexican markets today. Indigenous peoples in North America also practiced entomophagy. For instance, tribes in the USA and Canada consumed grasshoppers, crickets, caterpillars, flies, cicadas, beetles, ants, bees, and yellowjackets. The Tlicho people from Canada's Northwest Territories were known to eat warble fly larvae, which they collected during the butchering of caribou. In addition to their nutritional value, insects often held spiritual significance. For example, some cultures believed that eating wasps could bring prosperity, protection, and abundance. This tradition highlights the multifaceted role of insects in human societies, serving both as a food source and as a symbol of various cultural beliefs.

In recent years, edible insects have gained popularity among consumers, not only as snacks but also as side dishes and ingredients in various food products. Most of these insects fall into categories such as caterpillars, butterflies, moths, wasps, beetles, crickets, grasshoppers, bees, and ants. Studies on edible insects consistently highlight their impressive nutritional profile, particularly their high levels of proteins, lipids, fiber, and minerals [28]. This makes them an excellent substitute for traditional cereal flours, enhancing the nutritional value of snacks without compromising their technological properties. Sensory evaluations are generally positive, with only occasional changes in color noted. Incorporating insect-derived ingredients into meat products can enhance their mineral content and improve certain textural properties, such as elasticity. Additionally, insects serve as a viable alternative protein source for livestock feed, displaying favorable characteristics in extruded products. Insect protein isolates have shown bioactive properties, which can be beneficial in enhancing food formulations. These bioactive compounds include antioxidants, insulin regulators, and anti-inflammatory peptides, all of which contribute to high-value food products. Insect lipids, rich in essential fatty acids, have demonstrated positive effects on heart health, inflammation, and cancer prevention. The nutritional profile of these insects is making them valuable for nutritional enrichment in bakery products like cookies, bread, tortillas, and pasta. For example, adding 10% roasted speckled cockroach powder to wheat flour can significantly enhance the protein and fiber content of bread [31]. Similarly, replacing wheat flour with 5% insect flour from crickets, mealworms, or black soldier flies has shown to improve the nutritional profile and stability of bread dough. Additionally, cricket powder used in wheat bread boosts protein and essential amino acids, while mealworm powder enhances cereal-based snacks produced via 3D printing [31]. Corn tortillas fortified with 20% cricket protein powder also demonstrate improved nutritional quality without compromising taste or texture [30]. Companies are developing versatile insect-based ingredients, including soluble protein powders for beverages and textured proteins for meat substitutes. Insects are also being incorporated into meat analogues like burgers, meatballs, and sausages. Insect powders are used in protein supplements, energy bars, and beverages, with studies showing their potential in enhancing food formulations for infants and young children in developing countries, providing nutritional benefits and lower contamination risks compared to traditional blends. Innovative applications include using insect lipids in spreadable products like margarine, with up to 75% substitution possible without affecting quality [32]. Cookies made with insect oils have higher omega-3 fatty acids and vitamin E content, with high consumer acceptance. The development of novel processing methods is evident in the numerous patent applications for products such as cookies, rice cakes, energy bars, soup, tea, noodles, jerky, coffee, salad dressing, and tofu, reflecting a broad interest in incorporating edible insects into food and beverage formulations.

The burgeoning field of insect protein-based foods is being led by several pioneering companies, each contributing to the diversification and acceptance of this sustainable protein source [33]. Entomo Farms in Canada specializes in crickets and mealworms, offering products such as cricket powder and flavored snacks, which are prized for their high protein content and versatility [34]. Aspire Food Group, based in the USA, produces a wide array of cricket-based products including protein bars and whole roasted crickets, emphasizing nutritional benefits and environmental sustainability [35]. Jimini's, a European company, provides innovative insect-based items such as energy bars, granola, and pasta, catering to health-conscious and eco-friendly consumers [36]. Hargol FoodTech in Israel specializes in grasshopper protein powder, used in protein shakes and energy bars, highlighting the efficiency of grasshopper farming [37]. Six Foods in the USA produces Chirps, cricket-based chips marketed as a healthy snack alternative [38]. These companies are at the forefront of the insect protein movement, offering a range of products from bars and powders to whole insects and chips, thus addressing the increasing demand for sustainable and nutritious protein sources and positioning themselves as leaders in the transformation of global food systems.

Cultivated meat, also known as cultured meat, is authentic animal meat produced by cultivating animal cells directly, eliminating the need for traditional animal farming [39]. This process, which replicates the sensory and nutritional profiles of conventional meat by arranging the same cell types in similar structures, was first showcased by Dutch scientist Mark Post in 2013 with the unveiling of the first cultivated meat burger [40]. Since then, industry has expanded significantly, with over 150 companies across six continents and $2.6 billion in investments as of late 2022 [39]. The field has benefited from decades of research in cell culture, stem cell biology, tissue engineering, fermentation, and bioprocess engineering, with numerous companies and academic labs working to scale production. The process begins with acquiring and banking animal stem cells, which are then grown in bioreactors with an oxygen-rich cell culture medium containing nutrients and growth factors. These cells differentiate into muscle, fat, and connective tissues, which are then harvested and processed into final products, a process taking 2–8 weeks depending on the type of meat. Several companies are already transitioning to pilot-scale facilities, with regulatory approvals underway in various countries. The Singapore Food Agency approved the first cultivated chicken product in 2020, and in 2022, UPSIDE Foods completed a pre-market consultation with the FDA for its cultivated chicken [40]. Scaling up production will require solving challenges in cell lines, cell culture media, bioprocess design, scaffolding, and product design, necessitating substantial funding, new research initiatives, and a collaborative approach from existing and new companies in the life sciences.

6.3 Consumer's Acceptance and Adoption of Meat Substitutes

Interest in alternatives to meat products dates to the early nineteenth and twentieth centuries, primarily driven by economic factors [41]. Initially, vegetarian and vegan options were introduced, but these early products failed to replicate the taste and texture of meat satisfactorily. Over time, significant advancements have been made in developing plant-based meat substitutes, resulting in products that closely mimic the sensory experience of animal-derived meat. Despite the emergence of competitive alternatives like cultured meat and insect protein, both face technological and scaling challenges. Moreover, consumer acceptance remains a hurdle for these newer options. For instance, lab-grown beef, while being one of the most technologically advanced concepts, is still far from achieving commercial feasibility and is among the least favored meat substitutes. In the United States, a significant portion of respondents, about 74%, indicated a preference for conventional animal protein over cultured meat. Preferences for alternative proteins are often distinct and can be strongly correlated, yet each—plant-based meat substitutes, cultured meat, and edible insect protein—possesses unique sensory characteristics [42]. There is an influence of social companions (alone, friend, family, acquaintance, partner) and venue (home, cafe, bar, pub, food festival, restaurant) on the anticipated willingness to try a number of novel foods (insect-based foods, cultured meats, plant-based meat alternatives, and 3D printed foods) [43]. Eating with friends and at food festivals plays an important role in the anticipated acceptance of insect-based foods, cultured meats, and 3D printed foods in a similar way [43]. A systematic review [44] analyzed various alternative proteins and the factors influencing their acceptance by consumers. The alternative proteins examined include pulses (such as lentils and beans), algae, insects, plant-based meat alternatives, and cultured meat, each garnering varying levels of consumer acceptance. Key drivers of acceptance identified include motivations behind food choices, attitudes toward alternative protein sources, food neophobia or reluctance to try new foods, variations in protein preferences, and familiarity or exposure to these new market products.

Consumer acceptance of plant-based meat alternatives found that consumer behavior, influenced by complex socio-economic factors, plays a crucial role in the adoption of plant-based meat substitutes [41]. These alternatives were generally rated more favorably than other meat substitutes, although challenges related to sensory and nutritional qualities remain. While environmental and health benefits are important drivers for the market expansion of plant-based meat substitutes, they alone are not enough to ensure widespread adoption. Additionally, there is a lack of information on how hypothetical measures used in studies, such as willingness to pay, will translate into actual consumer behavior. Consumers engage in a variety of practices related to plant-based meat substitutes (PBMS), encompassing acquisition, preparation, and consumption. These practices are intertwined with a broader network that includes storage, substitution, food safety, and both meat-based and meat-free dietary habits, as well as plant-based and socio-cultural practices [45]. There is a significant impact of social and cultural frameworks in enhancing product awareness, developing skills, and imbuing meaning within the context of dietary and behavioral changes. The limited sensory appeal and low consumer awareness of the benefits of plant proteins contribute to modest market shares for meat substitutes [41]. A study [46] explored shifting consumer preferences from pork-based sausages to similar-looking plant-based counterparts by providing information about health and environmental impacts. Preferences were evaluated through blind tastings, packaging information, and informational dissemination. Initially, blind tastings favored meat products, but packaging information narrowed the preference gap. However, only additional health-related information equated consumer willingness to pay (WTP) and purchase preferences (PP) for both products. Similarly, combining environmental messages resulted in equal WTP, though PP still favored pork. This underscores the importance of disseminating detailed information on the benefits of plant-based proteins. Additionally, environmental, ethical, and health concerns have damaged meat's image for some consumers, increasing the relevance of meat substitutes (MS) and credence attributes labeling. However, it remains uncertain if MS match meat's nutritional standards and if Front-of-Package (FOP) labels reliably indicate quality [41]. Analysis of 5482 innovations in the German meat market from 2010 to 2018 using Mintel's Global New Product Database sees a rise in MS, peaking in 2015. Ofcom's A-score revealed MS contained fewer "nutrients to limit" and additives than red meat (RM) and poultry meat (PM). However, FOP labels did not consistently correlate with superior nutritional quality and fewer additives, highlighting the need for reliable information dissemination to influence consumer choices effectively [41]. Elzerman et al. 2015 explored the perceived suitability, appeal, intention to use, and sensory characteristics of meat substitutes in various dishes based solely on visual cues [47]. They developed an online survey to gauge consumer assessments of meat substitutes in different meal contexts. The study focused on six dishes: spaghetti, rice, wrap, pizza, pasta salad, and soup, evaluating their suitability for meat substitutes. The findings indicated that spaghetti, rice, and wraps were considered more suitable for meat substitutes than the other dishes. Combinations resembling typical Dutch meals, such as spaghetti with mince and rice with pieces, received the highest suitability ratings. These appropriateness ratings aligned with the attractiveness and intention scores. Another study [48] utilized open-ended questions, preference ranking, and identification tasks to investigate the sensory drivers and barriers to liking four types of burger patties: two plant-based burgers (one identified as a pea protein burger and another as an animal-like protein burger), a hybrid meat-mushroom burger, and a 100% beef burger. The study involved 175 untrained participants who were randomly assigned to either blind or informed conditions in a between-subject design. The primary objective was to assess the impact of information about the protein source/type (animal vs. plant-based) on consumer preferences and to gather product descriptors and levels of liking or disliking from participants. Results from the ranking tests indicated that the animal-like protein burger was the most preferred product in both blind and informed conditions, followed by the 100% beef burger. Interestingly, in the blind condition, no significant preference difference was observed between the beef burger, hybrid burger, and pea protein burger. Additionally, during the blind tasting, participants favored the pea protein burger over the hybrid burger, which contrasted with the results from the informed tasting. This suggests that factors beyond simple sensory enjoyment influence preferences. Furthermore, in the identification task, participants correctly identified the beef burger under the blind condition, yet still showed a preference for the plant-based animal-like burger.

These studies collectively demonstrate the potential for increased consumer acceptance of plant-based meat alternatives, while highlighting the complex interplay of sensory qualities, information dissemination, socio-economic influences, and social and cultural contexts in shaping consumer preferences and behaviors.

The rising demand for livestock products has led to numerous environmental, economic, and ethical concerns. To address these issues, alternative protein sources like edible insects have been developed, though consumer acceptance and commercialization present significant challenges. Several personal factors have emerged as significant predictors of consumer acceptance of insects as food, with various sociodemographic characteristics playing crucial roles. Gender was the most frequently examined characteristic, followed by age, education level, household income, and family size. For developing countries, information on gender and age was reported in only a few studies. Most studies found that men are more willing than women to accept insect-based products, regardless of the visibility of insects in the food. For example, an online survey in Italy revealed that men were 2.55 times more likely than women to try fried insects and meat burgers topped with larvae [49]. However, some studies did not find gender to be a significant factor. Age was also an important factor, with younger individuals generally showing more positive attitudes toward insect-based foods [41]. Younger people tend to be more aware of the environmental benefits of insect proteins and are less set in their dietary habits, making them more open to trying new foods. Conversely, studies in Japan and China found that older participants were more likely to consume insects due to their past experiences, highlighting the impact of established habits [41]. Nonetheless, some studies reported no significant correlation between age and the acceptance of insect-based foods [47]. Education level influences the willingness to consume insects, with more educated individuals often more inclined to try insect-based foods. For instance, higher education levels in Kenya were associated with a stronger intention to consume edible insects [41]. In Switzerland and Thailand, the impact of education on the acceptance of insect-based foods differed culturally. In Switzerland, more educated individuals were early adopters due to concerns about sustainability and health, while in Thailand, less educated individuals were more likely to consume insects, associating entomophagy with rural traditions. Household income and size also affect the acceptance of insect-based foods, though the influence varies by region. In China, higher household incomes and larger household sizes increased acceptance, while in Kenya, larger household sizes were associated with decreased acceptance. Acceptance rates can even vary within the same country, as seen in Italy and China, where respondents from different regions showed varying levels of willingness to accept insects as food [41]. Psychological factors, notably emotions and attitudes, significantly influence consumer acceptance of insects as food. Disgust emerges as a predominant negative determinant, particularly in developed nations. However, curiosity about the taste, texture, and novelty of insect-based foods can serve as a motivating factor for acceptance. Additionally, interest in the health benefits and environmental advantages of consuming insects, alongside a desire for dietary variety and novel food experiences, further enhances consumer willingness to embrace entomophagy. Food neophobia, characterized by a reluctance to try new foods, adversely affects the willingness to consume insect based products. Conversely, lower levels of neophobia are associated with increased acceptance. Similarly, food technology neophobia, which denotes apprehension toward foods produced using new technologies, also negatively impacts acceptance. Social pressures exert a considerable influence on consumer decisions, with both social norms and descriptive norms playing critical roles. Low social acceptance and negative social influences from family and peers can diminish the willingness to try insect-based foods [46]. In contrast, positive evaluations from peers and experts can enhance perceived quality and acceptance, fostering a more favorable attitude toward these products. The availability of insect-based products is another pivotal factor. When consumers perceive that these products are readily accessible, their propensity to purchase and consume them increases. Conversely, difficulty in obtaining these products can hinder the intention to incorporate them into their diet. Familiarity with insects, derived from cultural practices, exposure to entomophagy, or acquaintance with foreign cuisines, positively influences acceptance. In contrast, a lack of familiarity can result in the rejection of insect-based foods due to perceptions of insects as unnecessary or undesirable.

Research indicates that consumer preferences for insect species vary by region. In Italy, crickets, bee larvae, grasshoppers, mealworms, silkworms, and giant water bugs are favored, with adult insects being more acceptable than larvae. In Japan, wasp larvae and grasshoppers are preferred, while Romanian consumers favor locusts and ants. Sensory attributes such as appearance, taste, texture, and color are critical in shaping consumer acceptance. Consumers generally prefer insect-based products where insects are not visible. Taste is a significant predictor of acceptance, with positive taste experiences crucial for repeat consumption. Texture also plays a vital role; for instance, crispy textures are often preferred. Additionally, the perceived appropriateness of the product's carrier (e.g., meatballs vs. dairy drinks) can influence acceptance, with savory foods generally being more acceptable than sweet ones. Convenience and the perceived benefits of insect-based foods, such as their healthiness, nutritional value, and environmental sustainability, also affect consumer acceptance. Convenience, particularly in the form of snacks, can enhance readiness to accept insect-based foods. Perceived health and environmental benefits can increase willingness to try and consume these products, although conventional meat's perceived nutritional value can detract from this willingness. The novelty and safety of insect-based foods are also important. Novelty can deter acceptance due to preferences for traditional ingredients, while safety concerns related to contamination and disease can decrease willingness to consume and pay for insect-based products. Providing safety information can mitigate these concerns and positively influence acceptance.

Some studies reported that price significantly influences consumer decisions regarding insect-based foods and is often associated with perceived quality [49]. In developed countries, research shows mixed reactions to price variations. In the Netherlands, a study found that while the higher price of an insect-based burger (EUR 4) did not solely deter 64% of participants, it could, combined with other factors like taste and availability, impact future consumption [50]. Similarly, in Germany, higher prices positively influenced expectations and willingness to pay, associating cost with quality [51]. Conversely, lower prices reduced perceived quality and willingness to pay. Promotion tailored to consumer needs, emotions, and attitudes is essential for increasing acceptance. Effective communication about the benefits, safety, and risks of insect-based foods can enhance willingness to consume them. Highlighting health and sustainability benefits in promotional materials has been shown to attract positive consumer attention. In New Zealand, focus groups suggested promoting insect-based foods as convenient, healthy snacks at premium prices in sustainable packaging [49]. Transparent labeling about insect content and emphasizing the product's safety can also boost acceptance. Emotional appeals, such as promoting the health benefits of insect-based foods, are more effective than cognitive messages. However, some studies indicate that explicit nutritional claims, like high protein content, may decrease acceptance. Overall, strategic promotion emphasizing safety and benefits can positively influence consumer willingness to adopt insect-based foods.

The acceptance of cultured meat faces several barriers, many of which echo previous findings but now offer more detailed insights. A major issue remains the perceived unnaturalness of cultured meat, often underlying disgust and safety concerns. Despite efforts to address this, consumers remain unconvinced about its naturalness, with studies showing that those who value naturalness are more likely to reject cultured meat, although this concern varies geographically. For instance, European consumers show higher sensitivity to naturalness compared to Americans. Safety concerns, often tied to perceptions of unnaturalness, also persist, with consumers expressing anxiety about the long-term health effects of cultured meat. However, these concerns can be mitigated with additional information, as demonstrated by increased perceived safety when consumers are better informed. Transparency about health and safety is crucial for acceptance. Nutritional concerns also hinder acceptance, as consumers often doubt the nutritional value of cultured meat, perceiving it as inferior to conventional meat. This concern is linked to its artificiality, although it varies across demographics. For example, while Chinese consumers prioritize healthiness, German schoolchildren prefer cultured meat over insects despite health concerns. Trust issues encompass distrust in food companies, labeling, and regulatory bodies. Distrust in food scientists and conspiracy ideation predict rejection, and increasing trust in regulatory bodies and transparent communication can enhance acceptance. Disgust, related to perceived unnaturalness, is another strong barrier. Cultured meat is often seen as less disgusting than GMOs and insects but more so than synthetic additives. Disgust responses are emotional and can occur independently of rational evaluations, highlighting the need for familiarity to overcome this barrier. Food neophobia, the fear of novel foods, predicts rejection across various regions, with older consumers exhibiting a preference for familiar food preparations. Economic anxieties also hinder acceptance, with concerns about affordability and the broader economic impacts on farming and rural communities. There is anxiety about potential job losses in agriculture due to cultured meat production. Ethical concerns include potential harm to animals used for cell extraction and distrust in the industry's motives. Moral disgust, particularly regarding norm violations, significantly influences rejection. Acceptance is higher for cultured meat from traditionally consumed animals, with non-traditional species exacerbating ethical concerns. Addressing these barriers through targeted information, transparency, and trust-building measures is essential for increasing the acceptance of cultured meat. Price and sensory experience are critical determinants influencing consumer acceptance of cultured meat. Research underscores that price sensitivity is a significant barrier, with many consumers expressing concerns over the anticipated high cost of cultured meat compared to conventional options. Studies indicate varied expectations across different regions, where consumers in Brazil expect higher costs while those in the Dominican Republic anticipate lower prices. There's mixed willingness among consumers to pay a premium for cultured meat; while some are open to it, others remain price-sensitive, highlighting the need for affordability to attract mainstream adoption. Sensory experience, particularly taste and texture, also plays a pivotal role. Consumer expectations often lean toward perceiving cultured meat as inferior in taste and texture compared to traditional meat, which negatively impacts purchase intent. However, there's evidence suggesting that positive sensory experiences, such as tasting cultured meat perceived as comparable or even superior to conventional meat, can positively influence consumer acceptance. Addressing these factors through effective communication, sensory enhancement, and price competitiveness will be crucial for fostering broader consumer acceptance of cultured meat in the marketplace.

The exploration of alternative protein sources—mycoprotein, insect-based proteins, and cultured meat—reveals both promise and challenges in achieving widespread consumer acceptance. Mycoprotein stands out for its nutritional density, offering fiber, vitamins, and minerals comparable to traditional meat, yet its adoption requires overcoming perceptions of unfamiliarity. Insect-based proteins, historically significant across cultures, present robust nutritional profiles but encounter resistance due to Western aversions. Cultured meat, produced through cellular cultivation, holds potential to revolutionize meat production, yet faces substantial hurdles related to naturalness, safety concerns, and ethical implications.

Consumer acceptance hinges on several factors—perceived naturalness, safety, nutritional equivalence, sensory experience, and affordability. Addressing these concerns demands targeted strategies. Transparent communication about production methods and safety protocols is crucial to alleviate fears regarding novel technologies like cultured meat. Enhancing sensory attributes through research and development can bridge the taste and texture gap between alternative proteins and traditional meat, thereby enhancing consumer satisfaction. Moreover, affordability remains a barrier, necessitating cost-effective scaling of production methods to make alternative proteins competitive in the marketplace. Ethical considerations, particularly concerning animal welfare and environmental sustainability, require ongoing dialogue and adherence to rigorous standards. Collaborative efforts between industry, regulators, and consumer advocacy groups can foster trust and ensure alignment with evolving societal values. Regional differences in consumer preferences highlight the need for tailored marketing and educational campaigns that emphasize health benefits and environmental stewardship. The journey toward mainstream acceptance of alternative proteins is multifaceted, requiring innovation, regulatory support, and consumer education. By addressing concerns related to safety, naturalness, nutrition, sensory experience, affordability, and ethics, stakeholders can pave way for a sustainable food future that balances nutritional needs with environmental stewardship and ethical considerations.

References

1. Suri S, Ray S. Will a plant-based approach be sufficient to address food security, nutritional adequacy and sustainability in the future? Environ Urban ASIA. 2023;14(2):299–308.
2. Leroy F, Praet I, Meat traditions. The co-evolution of humans and meat. Appetite. 2015;90:200–11. https://doi.org/10.1016/j.appet.2015.03.014.
3. Leder D, Lehmann J, Milks A, Koddenberg T, Sietz M, Vogel M. The wooden artifacts from Schöningen's spear horizon and their place in human evolution. Proc Natl Acad Sci USA. 2024;121(15):e2320484121. https://doi.org/10.1073/pnas.2320484121.

4. Marín J, Saladié P, Rodríguez-Hidalgo A, Carbonell E. Neanderthal hunting strategies inferred from mortality profiles within the Abric Romaní sequence. PLoS One. 2017;12(11):e0186970. https://doi.org/10.1371/journal.pone.0186970.
5. Alt KW, Al-Ahmad A, Woelber JP. Nutrition and health in human evolution-past to present. Nutrients. 2022;14(17):3594. https://doi.org/10.3390/nu14173594.
6. Alexandratos N, Bruinsma J World agriculture towards 2030/2050: the 2012 revision. Agricultural Development Economics Division; 2012.
7. Kim J. An overview of Kenyan aquaculture: current status, challenges, and opportunities for future development. Fish Aquat Sci. 2014;17(1):1–11. https://doi.org/10.5657/FAS.2014.0001.
8. Zhang Z, Liu H, Jin J, Zhu X, Han D, Xie S. Towards a low-carbon footprint: current status and prospects for aquaculture. Water Biol Secur. 2024;26:100290. https://doi.org/10.1016/j.watbs.2024.100290.
9. Delgado CL. Rising consumption of Meat and Milk in developing countries has created a new food revolution. J Nutr. 2003;133(11):3907S–10S. https://doi.org/10.1093/jn/133.11.3907S.
10. Fukase E, Martin W. Economic growth, convergence, and world food demand and supply. World Dev. 2020;132:104954. https://doi.org/10.1016/j.worlddev.2020.104954.
11. Leitzmann C. Vegetarian nutrition: past, present, future. Am J Clin Nutr. 2014;100(Suppl 1):496S–502S. https://doi.org/10.3945/ajcn.113.071365.
12. Arora S, Kataria P, Nautiyal M, Tuteja I, Sharma V, Ahmad F. Comprehensive review on the role of plant protein as a possible Meat analogue: framing the future of Meat. ACS Omega. 2023;8(26):23305–19. https://doi.org/10.1021/acsomega.3c01373.
13. Bakhsh A, Park J, Baritugo KA, Kim B, Sil MS. A holistic approach toward development of plant-based meat alternatives through incorporation of novel microalgae-based ingredients. Front Nutr. 2023;10:1110613. https://doi.org/10.3389/fnut.2023.1110613.
14. Rubio NR, Xiang N, Kaplan DL. Plant-based and cell-based approaches to meat production. Nat Commun. 2020;11(1):6276. https://doi.org/10.1038/s41467-020-20061-y.
15. Safdar B, Zhou H, Li H, Cao J, Zhang T, Ying Z. Prospects for plant-based Meat: current standing, consumer perceptions, and shifting trends. Foods. 2022;11(23):3770. https://doi.org/10.3390/foods11233770.
16. Gomes A, Sobral PJDA. Plant protein-based delivery systems: an emerging approach for increasing the efficacy of lipophilic bioactive compounds. Molecules. 2021;27(1):60. https://doi.org/10.3390/molecules27010060.
17. The Rich History of Tofu: From Ancient Origins to Modern Delicacy, Retrieved July 4, 2024, from https://nihaoma-mandarin.com/culture/the-rich-history-of-tofu-in-china-from-ancient-origins-to-modern-delicacy/.
18. Alam MS, Kaur J, Khaira H, Gupta K. Extrusion and extruded products: changes in quality attributes as affected by extrusion process parameters: a review. Crit Rev Food Sci Nutr. 2016;56(3):445–73. https://doi.org/10.1080/10408398.2013.779568.
19. A Journey into Tempeh – Foodies. Retrieved July 4, 2024, from https://foodies.id/a-journey-into-tempeh/.
20. Flambeau M, Redl A, Respondek F. Proteins from wheat: sustainable production and new developments in nutrition-based and functional applications. Sustain Prot Sour. 2016:67–78. https://doi.org/10.1016/B978-0-12-802778-3.00004-4.
21. Reynaud Y, Buffière C, Cohade B, Vauris M, Liebermann K, Hafnaoui N. True ileal amino acid digestibility and digestible indispensable amino acid scores (DIAASs) of plant-based protein foods. Food Chem. 2021:338. https://doi.org/10.1016/j.foodchem.2020.128020.
22. What is plant-based meat? | The Good Food Institute. Retrieved July 5, 2024, from https://gfi.org/plant-based/.
23. Berka RM, Nelson BA, Zaretsky EJ, Yoder WT, Rey MW. Genomics of Fusarium venenatum: an alternative fungal host for making enzymes. Appl Microbiol Biotechnol. 2003;4:191–203. https://doi.org/10.1016/S1874-5334(04)80010-8.
24. Majumder R, Miatur S, Saha A, Hossain S. Mycoprotein: production and nutritional aspects: a review. Sustain food technol. 2004;2(1):81–91. https://doi.org/10.1039/D3FB00169E.

25. Promyc® | Promyc – the next-level mycoprotein ingredient fungi-based alternative protein for the food industry | Mycoprotein | Sweden. Retrieved July 5, 2024, from https://mycorena.com/promyc.
26. Derbyshire EJ, Delange J. Fungal protein – what is it and what is the health evidence? A systematic review focusing on Mycoprotein. Front Sustain Food Syst. 2021:5. https://doi.org/10.3389/FSUFS.2021.581682/FULL.
27. John DA, Babu GR. Lessons from the aftermaths of green revolution on food system and health. Front Sustain Food Syst. 2021:5. https://doi.org/10.3389/FSUFS.2021.644559/FULL.
28. Olivadese M, Dindo ML. Edible insects: a historical and cultural perspective on entomophagy with a focus on Western societies. Insects. 2023;14(8):690. https://doi.org/10.3390/insects14080690.
29. Kelhoffer JA. Did John the Baptist eat like a former Essene? Locust-eating in the ancient near east and at Qumran. JSTOR. 2004;11(3):293–314.
30. Liceaga AM. Edible insects, a valuable protein source from ancient to modern times. Adv Food Nutr Res. 2022;101:129–52. https://doi.org/10.1016/bs.afnr.2022.04.002.
31. Gantner M, Sadowska A, Piotrowska A, Kulik K, Sionek B, Kostyra E. Wheat bread enriched with House cricket powder (Acheta domesticus L.) as an alternative protein source. Molecules. 2024;29(3):711. https://doi.org/10.3390/molecules29030711.
32. Smetana S, Leonhardt L, Kauppi S, Pajic A, Heinz V. Insect margarine: processing, sustainability and design. J Clean Prod. 2020;264:121670. https://doi.org/10.1016/j.jclepro.2020.121670.
33. Lisboa HM, Nascimento A, Arruda A, Sarinho A, Lima J, Batista L, Dantas MF, Andrade R. Unlocking the potential of insect-based proteins: sustainable solutions for global food security and nutrition. Foods. 2024;13(12):1846. https://doi.org/10.3390/foods13121846.
34. Home | Entomo Farms. Retrieved July 5, 2024, from https://entomofarms.com/.
35. Cricket Protein Farming For Cricket Powder Production – Aspire Food Group. Retrieved July 5, 2024, from https://aspirefg.com/.
36. My First Insect – Jimini's. Retrieved July 5, 2024, from https://www.jiminis.com/en/blog/my-first-edible-insect-jimini/.
37. Home-new – Hargol FoodTech. Retrieved July 5, 2024, from https://hargol.com/.
38. Chips Made with Cricket Flour – As Seen on Shark Tank | Chirps Chips. Retrieved July 5, 2024, from https://eatchirps.com/.
39. Is consuming lab-grown meat ethical, nutritious, safe? What the evidence says. Retrieved July 5, 2024, from https://www.orfonline.org/research/is-consuming-lab-grown-meat-ethical-nutritious-safe.
40. Wang Y, Zou L, Liu W, Chen X. An overview of recent progress in engineering three-dimensional scaffolds for cultured Meat production. Foods. 2023;12(13):2614. https://doi.org/10.3390/foods12132614.
41. Szenderák J, Fróna D, Rákos M. Consumer acceptance of plant-based Meat substitutes: a narrative review. Foods. 2022;11(9):1274. https://doi.org/10.3390/foods11091274.
42. Lima M, Costa R, Rodrigues I, Lameiras J, Botelho G. A narrative review of alternative protein sources: highlights on Meat, fish, egg and dairy analogues. Foods. 2022;11(14):2053. https://doi.org/10.3390/foods11142053.
43. Motoki K, Park J, Spence C, Velasco C. Contextual acceptance of novel and unfamiliar foods: insects, cultured meat, plant-based meat alternatives, and 3D printed foods. Food Qual Prefer. 2022;96:104368. https://doi.org/10.1016/j.foodqual.2021.104368.
44. Onwezen MC, Bouwman EP, Reinders MJ, Dagevos H. A systematic review on consumer acceptance of alternative proteins: pulses, algae, insects, plant-based meat alternatives, and cultured meat. Appetite. 2021;159:105058. https://doi.org/10.1016/j.appet.2020.105058.
45. White SK, Ballantine PW, Ozanne LK. Consumer adoption of plant-based meat substitutes: a network of social practices. Appetite. 2022;175:106037. https://doi.org/10.1016/j.appet.2022.106037.
46. Martin C, Lange C, Marette. Importance of additional information, as a complement to information coming from packaging, to promote meat substitutes: a case study on a sausage

based on vegetable proteins. Food Qual Prefer. 2020;87:104058. https://doi.org/10.1016/j.foodqual.2020.104058.
47. Elzerman JE, Hoek AC, van Boekel T, Luning PA. Appropriateness, acceptance and sensory preferences based on visual information: a web-based survey on meat substitutes in a meal context. Food Qual Prefer. 2015;42:56–65. https://doi.org/10.1016/j.foodqual.2015.01.010.
48. Fiorentini M, Kinchla AJ, Nolden AA. Role of sensory evaluation in consumer acceptance of plant-based Meat Analogs and Meat extenders: a scoping review. Foods. 2020;9(9):1334. https://doi.org/10.3390/foods9091334.
49. Alhujaili A, Nocella G, Macready A. Insects as food: consumers' acceptance and marketing. Foods. 2023;12(4):886. https://doi.org/10.3390/foods12040886.
50. House J. Consumer acceptance of insect-based foods in The Netherlands: academic and commercial implications. Appetite. 2016;107:47–58. https://doi.org/10.1016/j.appet.2016.07.023.
51. Kornher L, Schellhorn M, Vetter S. Disgusting or innovative-consumer willingness to pay for insect based burger patties in Germany. Sustainability. 2018;11(7):1878. https://doi.org/10.3390/su11071878.

Chapter 7
Meat Alternatives and Its Potential Global Health Impact

The global food system faces a great challenge in order to meet the dietary needs of a growing population while reducing its impact on the negative health outcomes. To address these challenges, both cell-based and plant-based products are emerging as potential substitutes. It is widely acknowledged that animal-based products typically have a higher greenhouse gas (GHG) emissions intensity compared to plant-based food commodities. However, a recent study by the Global Burden of Diseases (GBD) reveals that red and processed meat, in particular, are quantified risk factors for non-communicable diseases (NCDs).

In light of global health concerns and the sustainable development goals (SDGs), particularly SDG 2 (zero hunger) and SDG 13 (action to combat climate change), there is a pressing need to align meat consumption with efforts to address climate conditions.

This is crucial not only for reducing the risk of NCDs associated with red and processed meat consumption but also for achieving global targets on NCDs as part of SDG 3 (good health and well-being). Nevertheless, meat substitutes have the potential to make substantial contributions to reducing GHG emissions, thereby benefiting global health. This underscores the importance of further research and policy measures to promote the adoption of meat substitutes as part of a comprehensive strategy to address both climate change and public health concerns.[1]

When examining the issue of "animal slaughter," it becomes apparent that there are three significant concerns intertwined with this practice. Firstly, there are ecological concerns, as the production of traditional meat has been shown to damage the environment through the consumption of large amounts of land, water, and the production of substantial greenhouse gas (GHG) emissions. Secondly, there are

[1]Ritchie H, Reay DS and Higgins P (2018) Potential of Meat Substitutes for Climate Change Mitigation and Improved Human Health in High-Income Markets. Front. Sustain. Food Syst. 2:16. https://doi.org/10.3389/fsufs.2018.00016

S. Ray, S. Suri, *The Sustainable Plate: Unveiling the Science of Meat Substitutes and Impact on Global Health*, https://doi.org/10.1007/978-3-031-94604-2_7

personal health concerns, as the consumption of traditional meat has been linked to adverse health effects, including an increased risk of cardiovascular disease and cancer, leading to higher morbidity rates. Thirdly, there are ethical concerns, as the mere act of killing animals for human consumption raises ethical dilemmas for many.

Despite these concerns, eradicating the consumption of meat proves challenging due to its widespread popularity. Recognizing a business opportunity, the food industry has responded by introducing meat substitutes, such as plant-based meat alternatives (PBMA) and lab-grown cultured meat (CM). While PBMA, a well-textured vegetable protein, initially faced criticism for its taste, flavor, and texture, the development of PBMA 2.0 has resulted in products that closely resemble traditional meat in appearance, taste, and flavor.

The comparative environmental impact of traditional meat versus PBMA and CM is striking. The latter require considerably less land and water and emit fewer GHGs during production. Additionally, studies suggest that young university students are more receptive to trying plant-based alternatives over traditional meat. However, there remains a preference for traditional meat in certain markets, as evidenced by consumer behavior in blind taste tests. Participants in France showed a preference for pork-based sausages over plant-based alternatives, while participants in the USA expressed a willingness to pay more for 100% beef burgers.

Despite the potential benefits of PBMA and CM for the environment, health, and animal welfare, challenges remain. Cost is a significant factor, as plant-based meat is often priced at a premium compared to conventional beef. Additionally, the availability of meat substitutes is still relatively limited compared to traditional meat options. While the demand for meat is unlikely to diminish and may even increase, the shift toward plant-based alternatives presents an opportunity to address environmental, health, and ethical concerns associated with traditional meat consumption. Continued research and innovation in the field of meat substitutes are essential to meet the growing demand for sustainable and ethical food options.[2]

Protein plays a vital role in every body function and is a fundamental building block of life. In total, 75% of dry weight in human tissue comprises of protein. It is crucial for enzyme functions, immune responses, hormonal regulations, nutrition transport, muscle building, and repair.

Protein energy malnutrition is referred to insufficient caloric and protein intake. According to Our World in Data, approximately 1,89,345 people died globally in the year 2021 due to protein energy malnutrition. Food and agriculture organizations also estimated that more than 2 billion people in the world experience moderate to severe food insecurity which is also contributed by reduced protein intake.

Red meat and processed meat consumption, although a rich source of protein, has been observed to have potential impact on chronic diseases like heart disease, certain cancers, and diabetes. This has greatly influenced people globally to find healthy alternatives. A study conducted by California Seventh Day Adventists

[2] John B Nezlek, Catherine A Forestell, "Meat substitutes: current status, potential benefits, and remaining challenges", Current Opinion in Food Science, Volume 47, 2022, ISSN 2214–7993, https://doi.org/10.1016/j.cofs.2022.100890

among 25,153 participants showed that regular consumption of meat was associated with 70% and 37% higher risk in men and women, respectively, to get fatal ischemic heart disease. Another research by the World Cancer Research Fund (2007) highlighted the relationship between red meat consumption and the risk of colorectal cancer where approximately 34,000 cases per year of colorectal cancer are reported mainly due to meat consumption.

A 2023 report in The American Journal of Clinical Nutrition presented a prospective study on red meat intake, using a semi-quantitative food frequency questionnaire, that showed a clear link between increased risk of type 2 diabetes and red meat consumption. Compared to those who ate the least red meat, people who consumed more red meat had 62% higher risk of developing type 2 diabetes. With respect to processed meat, the risk was 51% higher. Interestingly, substituting nuts and legumes with red meat was associated with 30% reduction in type 2 diabetes.

Along with these there are other pitfalls in health like high saturated fat content in red meat leading to high cholesterol levels, potentially increasing the risk of heart diseases and also certain inflammatory bowel diseases. In contrast to the above, as an alternative to red and processed meat, several studies have observed rising interest in plant-based and cell-based protein products. This is likely to fuel the growing awareness on health benefits associated with meat alternative diets.

Plant-based alternatives have shown to be a good source of protein, fiber, and essential nutrients. Substituting red meat with plant-based and cell-based protein can contribute to lower risks of chronic diseases and meet the nutrition requirement of an individual especially when consumed with balanced diet. Plant-based meat substitutes like soya protein, pea protein, and mycoprotein have shown that they offer similar textures and flavors like meat, but with low saturated fat content and high fiber levels. Cell-based meat uses animal cells grown in laboratory environments at early stages of development but shows a potential to offer more bioequivalent alternative to conventional meat. The surge of these products is mostly driven by environmental concerns, health concerns, and advancements in matching the taste and texture of traditional red meats and processed meats.

A study done by Meier et al. in 2018 that used food demand elasticity models suggested that substituting meat by plant-based alternative with "Flexitarian" dietary approach, where people primarily consume plants based with occasional meat consumption, could potentially prevent an estimated 8.1 million premature deaths globally per year by 2050 due to a decrease in diet-related chronic diseases. This translates to a potential 5.5 million lives saved annually just by replacing 10% of red meat consumption with plant-based protein sources.

A 2015 report by the EAT-Lancet Commission proposed a global dietary framework for healthy and sustainable diets. This framework emphasized the reduction in red meat consumption and increased intake of fruits, vegetables, legumes, and whole grains. They estimated that adopting such dietary patterns could prevent approximately 10.9 million deaths annually by 2050 due to a decrease in noncommunicable diseases. Recent publication from NIH-AARP diet and health study did a large, prospective cohort to examine the impact of plant-based protein choice on mortality. In total, 6,17,199 participants aged between 50 and 71 from the USA were followed from 1995 till 2011. The study found that substituting just 3% of animal protein intake with plant protein was associated with a 10% reduction in overall mortality.

Research suggests that protein-based proteins can help with the improvement of gut health. A study report by *Nature Microbiology* (2017) found that proteins from plant sources increased the healthy gut bacteria which helped with the reduction of gut inflammation and improved the overall immune system. Another study published by *The American Journal of Clinical Nutrition* (2017) pointed out that individuals who consumed more plant-based protein diet had significantly lower levels of LDL cholesterol that is usually linked to cardiac diseases.

The shift toward plant-based meat alternatives (PBMA) and cultured meat (CM) presents a crucial opportunity to address environmental, health, and ethical concerns associated with traditional meat consumption. The higher greenhouse gas emissions of animal-based products, particularly red and processed meat, underscore the urgent need for sustainable dietary practices in line with global health concerns and sustainable development goals (SDGs) such as SDG 2 (zero hunger), SDG 13 (climate action), and SDG 3 (good health and well-being). The environmental benefits of PBMA and CM are significant, requiring less land and water and emitting fewer greenhouse gases compared to conventional meat production. This transition not only supports climate change mitigation efforts but also reduces the risk of non-communicable diseases (NCDs) linked to red and processed meat consumption.

Despite the potential benefits, widespread adoption of meat substitutes faces challenges, primarily due to higher costs and limited availability. The food industry's development of PBMA 2.0, which closely resembles traditional meat in appearance, taste, and flavor, marks progress in overcoming consumer resistance. However, achieving significant market penetration will require further research and policy measures to make these products more affordable and accessible. Ethical concerns regarding animal slaughter, alongside ecological and personal health issues, further justify the promotion of meat substitutes. Younger consumers, particularly university students, show a higher receptivity to plant-based alternatives, although traditional meat remains preferred in certain markets. Overall, while traditional meat consumption is unlikely to diminish significantly, the growing demand for sustainable and ethical food options necessitates continued innovation and support for meat substitutes. This approach is vital for achieving long-term sustainability goals and improving global health outcomes.

In common parlance, food security is contrary to the concept of food insecurity. Expanding further, according to the World Bank, there are four dimensions: (i) availability, (ii) economic and physical access, (iii) nutritional value, and (iv) long-term accessibility.[3] Presently, 691 to 783 million people in the world are facing the

[3] What is food security? https://www.worldbank.org/en/topic/agriculture/brief/food-security-update/what-is-food-security

hunger problem.[4] The World Bank is one of the leading financiers with US$ 9.6 million in 2022, concurrently promoting climate-smart techniques to enhance the food system.

PBMA/CM represents two primary branches of meat substitutes within the meat substitute industry. These substitutes consist of water, vegetables, additives, coloring agents, binding agents such as soy, egg white, and other adhesive agents. One of the important influencing factors for consumers of PBMA/CM are appearance, taste, texture, smell, and experience of the product. We can safely say emulation is the key to attracting potential consumers. The marketability of the product heavily relies on appearance, creating a positive perception in the mind of potential consumers. When asked why meat emulation is significant, the simple answer is meat substitutes face a greater challenge in winning over conventional meat and converting traditional meat eaters.[5]

Men tend to prefer conventional meat, and women are more inclined toward substitutes. This factor is dynamic and subjective, potentially having a diminishing effect over time.[6] Some findings do indicate that a younger group are more inclined toward alternatives in comparison to their counterparts. Further subdivision reveals three sub-classifications of consumers under the former, i.e., (i) omnivores, (ii) pescetarians, and (iii) flexitarians. The preferences of these categories have a considerable impact on the meat substitute industry.

Praxis of "occasion" used here is the inclusive context, not limiting it to some particular time. Individuals consume multiple meals throughout the day, and tracking them based on patterns and frequency provides a comprehensive understanding of how and where these substitutes may be utilized. Occasions such as Sunday/weekday meals with family, alone, with friends, dinner invitations, business meetings, and barbeque parties offer opportunities for individuals to try these substitutes. Frequency preferences may vary, ranging from weekly to abstaining during business or family gatherings. Omnivores find meat substitutes equally "appropriate" on all listed occasions.

In contrast, flexitarians prefer them most with family on a weekday as well as alone. On the other hand, they find it "very inappropriate" during socializing occasions. Meat and meat substitutes, such as nuggets, steaks, and sausages, are often perceived equally in terms of taste and visual appeal. Interestingly, consumers prefer meat substitutes that do not taste like meat but desire them to resemble meat in appearance, texture, and perfection. Overall, omnivores prefer meat over meat substitutes, while flexitarians also favor meat except for consideration of nutrient value and environmental friendliness. However, there remains a significant gap that needs

[4] Food security and nutrition situation remains dire in 2022, https://www.un.org/en/global-issues/food#:~:text=According%20to%20the%202023%20edition,million%20people%20compared%20to%202019

[5] Replacement of meat by meat substitutes. A survey on person- and product-related factors in consumer acceptance https://www.sciencedirect.com/science/article/abs/pii/S0195666311000523?via%3Dihub

[6] Consumers' associations, perceptions and acceptance of meat and plant-based meat alternatives https://doi.org/10.1016/j.foodqual.2020.104063

to be addressed to increase both potential acceptability and physical availability of meat substitutes. The research in the domain of meat substitutes is expanding, with primary focus on the aspect of affordability, nutrient values, perception building, acceptability, and future opportunity.[7] The meat substitutes market is projected to grow until 2027 at USD 15.7 billion from USD 7.9 in 2022,[8] while the meat industry is estimated to grow at 5x times that rate.

Among the two significant barriers for PBMA/CM, "price" ranks second after "taste." However, with these challenges, there is a promising element such as steady growth in the market share, expected to reach USD 15.7 billion by 2027, which is something that can be looked upon to reduce prices. This growth not only anticipates a positive shift toward reducing prices to compete more effectively with the conventional meat industry but also signifies a proliferating acceptance of plant-based alternatives among consumers globally.[9] According to a comprehensive survey comparing the prices of plant-based meats such as chicken, pork, and beef, they are respectively four times more expensive than chicken, three times more expensive than pork, and two times more expensive than beef.[10] It is important to highlight the current price differentials, which must be addressed in order to accelerate the adoption of meat substitutes.

Observations derived from the US retail prices shed light on two pivotal points. Firstly, the retail price of plant-based beef is closely comparable to conventional beef. Secondly, achieving the sweet spot price is feasible by drawing insights from private labels. This understanding is highlighted by contrasting the prices of plant-based beef and conventional beef in the US retail market. Plant-based beef brands such as Ozo Grounds and Beyond Beef are priced at $10.65/lb. and $9.99/lb., respectively. In contrast, conventional brands like Albertsons Open Nature Grass-Fed Angus beef grounds and Kroger ground beef are listed at $7.49/lb. and $5.59/lb., respectively. Additionally, the rapid growth of private-label plant-based meats, which surged by 72% in 2019, signifies a noteworthy trend. This trend reflects a consumer preference for private-label products due to their comparative affordability when juxtaposed with branded counterparts. Consequently, there exists a compelling case for market players to heed these dynamics and strategize accordingly to tap into the burgeoning demand for plant-based alternatives while maintaining comparative pricing structures.

Is the element of "price" the paramount consideration for consumers, or do other ancillary factors play a significant role in determining the fate of PBMA/CM products? Consumer intent is influenced not only by taste but also by product pricing.

---

[7] Consumers' associations, perceptions and acceptance of meat and plant-based meat alternatives https://doi.org/10.1016/j.foodqual.2020.104063

[8] https://www.marketsandmarkets.com/Market-Reports/plant-based-meat-market-44922705.html

[9] Reducing the price of alternative proteins https://gfi.org/wp-content/uploads/2021/12/Reducing-the-price-of-alternative-proteins_GFI_2022.pdf

[10] 2020 Retail Meat Review: Robust Sales With Plant-Based Products Gaining Ground https://progressivegrocer.com/2020-retail-meat-review-robust-sales-plant-based-products-gaining-ground

While health and nutritional values hold a prominent position on the consumer's priority checklist, these aspects will be addressed in the subsequent segment of the literature review.[11]

Research indicates that adults who do not consume PBMA/CM exhibit a higher preference (66%) for conventional meat, with only a few (20%) considering it expensive. Even among occasional PBMA/CM consumers, a significant portion (50%) still prefer conventional meat, with only 26% perceiving PBMA/CM as expensive. These results are intriguing, as they indicate that less than 50% perceive PBMA/CM as expensive, suggesting potential biases toward conventional meat. Despite the comparatively higher prices, 23% still find PBMA/CM expensive.

Consumers generally demonstrate a willingness to pay when the price is the same as that of conventional meat, which could explain the high preference rates observed earlier. Early adopters, such as flexitarians (46%) and millennials (45%), represent potential target audiences willing to pay more for PBMA/CM. The sweet spot price range, typically 5–10% higher, can be achieved through effective promotional activities, ensuring consumer acceptance and market growth over time.[12]

In countries such as Australia, Germany, Japan, Italy, Sweden, the UK, and the USA, retail prices are indeed high, yet they remain within reach. Product pricing plays a crucial role in driving acceptance, raising awareness, fostering willingness, and influencing the preferences of meat eaters. This can be achieved by drawing insights into private-label plant-based products, which offer valuable lessons for market positioning. Simultaneously, the taste of the product is a significant concern for consumers, along with cultural and traditional preferences regarding meat. Nutrient value also ranks as one of the key factors influencing the consumer's priority checklist, highlighting the multifaceted nature of consumer decision-making in the realm of plant-based meat alternatives.[13]

Retail prices are high yet achievable in countries like Australia, Germany, Japan, Italy, Sweden, UK, and the USA. Pricing of the product has a major role in paying for acceptance, awareness, willingness, and converting preference of meat eaters. And the same is achieved through the insights from the private label plant-based products. At the same time, taste of the product is something that bothers the consumer along with cultural and traditional preference toward meat. Nutrient value is also one of the key factors that affect the priority checklist of the consumer.

According to the International Vegetarian Union, close to 600 million people worldwide have shifted toward a plant-based diet. Alongside this surge in popularity, there is an increased availability of meat substitutes. Brands like Beyond Meat, Impossible Foods, Eat JUST, 'Kellogg's Incogmeato, and MorningStar Farms, and 'Maple Leaf Foods' are some of the top global brands available in the retail market.

---

[11] Reducing the price of alternative proteins https://gfi.org/wp-content/uploads/2021/12/Reducing-the-price-of-alternative-proteins_GFI_2022.pdf

[12] US Plant-based Proteins Market Report 2023 https://store.mintel.com/report/us-plant-based-proteins-market-report

[13] Reducing the price of alternative proteins https://gfi.org/wp-content/uploads/2021/12/Reducing-the-price-of-alternative-proteins_GFI_2022.pdf

This dietary shift is primarily driven by concerns for animal welfare, environmental friendliness, and health.

Meat substitutes found in retail markets typically contain ingredients such as water, soy, wheat, egg whites, whey protein, oils, binding agents, and coloring agents. These ingredients are processed to emulate the taste, texture, appearance, ease of preparation, and nutrient profile of conventional meat. However, research indicates that such processing can lead to "ultra-processing," a term widely acknowledged despite lacking international consensus.

Excessive consumption of these processed foods is associated with an increased risk of non-communicable diseases, including overweight, obesity, type 2 diabetes, cardiovascular diseases, and cancer.[14] On the other hand, the need for meat substitutes has arisen from excessive red and processed meat leading to type 2 diabetes, colorectal cancer, and reduced life expectancy.[15] In such a case, a plant-based diet is known for maintaining balance if consumed at an accurate marker of processing and wisely.

This raises important questions about the appropriate level of processing required to ensure that meat substitutes remain a healthy and nutritious option. This would be achieved by studying the nutrient profile as well as the degree of processing.

It is not always the case that excluding conventional meat would automatically result in a healthier option. The research discussed here is one of the very few to conduct a comprehensive two-dimensional study on plant-based alternative products (PBAP), encompassing both the nutritional and health (processing) impacts. PBMA products available in the Spanish market were thoroughly evaluated. Information regarding these products was sourced from three primary sources provided by the manufacturer: (i) fact sheets, (ii) official website, or (iii) packaging labels. This meticulous examination sheds light on the complexities surrounding the health implications of plant-based meat alternatives.

Egg and seafood-based plant alternatives were found to have low protein and saturated fat compared to their animal counterparts. Similarly, other PBMA displayed lower protein and saturated fat but higher fiber and salt content than conventional meat. Most PBMA products fell into the top three categories (A, B, and C) of the nutritional value rating. In terms of processing, these PBMA were classified as "ultra-processed" due to efforts to emulate sensory elements. Coloring and additives, such as texture enhancers, thickeners, gelling agents, carbohydrates, and oil emulsions, were utilized to enhance appearance and mimic conventional meat.

Food security, contrary to food insecurity, encompasses four dimensions: availability, economic and physical access, nutritional value, and long-term accessibility. With a significant portion of the global population facing hunger, the World Bank has invested in climate-smart techniques to improve the food system. The physical availability of plant-based meat alternatives (PBMA) and cultivated meat (CM) is

---

[14] Are plant-based alternatives healthier? A two-dimensional evaluation from nutritional and processing standpoints https://www.sciencedirect.com/science/article/pii/S0963996923004027

[15] Novel plant-based meat alternatives: future opportunities and health considerations https://pubmed.ncbi.nlm.nih.gov/36603854/

influenced by factors like appearance, taste, texture, and smell. These elements are crucial for emulating traditional meat to attract conventional meat eaters. Consumer acceptance of PBMA/CM is driven by gender and consumption occasions, with women and younger consumers showing a higher preference for these substitutes. Different consumer categories, such as omnivores, pescetarians, and flexitarians, have varying preferences impacting the meat substitute industry.

Affordability remains a major barrier for PBMA/CM, with prices significantly higher than conventional meat. However, market growth projections suggest potential for reduced prices and increased acceptance. Price comparisons indicate that private-label products could play a significant role in achieving competitive pricing. A consumer's willingness to pay is influenced by both taste and price, with early adopters like flexitarians and millennials showing a readiness to pay more.

Nutritional value is another critical factor, with a global shift toward plant-based diets driven by concerns for health, animal welfare, and the environment. Despite the nutritional benefits, the ultra-processing of PBMA raises health concerns, such as increased risks of non-communicable diseases. Studies indicate that while PBMA products often have lower protein and saturated fat compared to conventional meat, they may also contain higher fiber and salt levels.

Overall, the market for meat substitutes is projected to grow significantly, with opportunities to address affordability, perception, and nutritional concerns. Enhancing the taste, texture, and nutritional profile of PBMA/CM, while managing processing levels, will be crucial for increasing consumer acceptance and establishing a sustainable food system. While assimilating the information on PBMA/CM, one is bound to infer aspects of a sustainable diet, which leads to questions about how these alternatives perform from a social equity and food justice perspective. However, before delving into this latter aspect, it is important to discuss the sustainability of PBMA/CM considering dietary preferences.

7.1 Meaning of Sustainable Diets?

The Food and Agriculture Organisation of the United Nations theorized the definition of Sustainable Diet[16]: *"Sustainable Diets are those diets with low environmental impacts which contribute to food and nutrition security and to healthy life for present and future generations. Sustainable diets are protective and respectful of biodiversity and ecosystems, culturally acceptable, accessible, economically fair and affordable; nutritionally adequate, safe and healthy; while optimizing natural and human resources"*.[17] The concept of sustainable diet takes into consideration the present as well as the future needs, where the food needs to be nutritious,

[16] Sustainability of plant-based diets: back to the future https://www.sciencedirect.com/science/article/pii/S0002916523048992

[17] Burlingame B, Dernini S, eds. Sustainable diets and biodiversity: directions and solutions for policy, research and action. Proceedings of the International Scientific Symposium. Biodiversity

economically affordable, accessible, culturally acceptable, and ecologically respectful. It pushes the argument one step ahead where one has to be mindful when it comes to utilization of resources and ecological degradation. At the same time it is imperative to understand that the definition varies within subjects such as it would not be the same for the consumer as it would be for the farmer or manufacturer.

7.2 Shifting Patterns: Agricultural Produce

The food system thrives on two factors—the input and output. The input consists of fossil fuel, chemicals, nitrogen fertilizers, and irrigation water, (mostly non-renewable resources) which are essential elements for agricultural produce. On the other hand, the output consists of the waste emission and final product that is developed, i.e., the product. Whenever the balance between the two is disrupted, it results in inevitable ecological degradation. Researchers put the burden on the element of "societal demand," which is a potential disruptor, and the allied practices, which becomes necessary to meet this increasing demand.

The shift in agricultural practices has been significant as the societal demand increases. Historically, a single tract of land supported both crop cultivation and livestock rearing through integrated polyculture practices. The resources used for the production of agricultural produce were low when compared with the output derived. But with looming industrial agriculture, there is a shift from polyculture to monoculture. This has led to increased resource use (inputs), resulting in an energy imbalance. Thus, the shifting pattern in agriculture has a significant impact on the balance between utilization of resources and ecological degradation.

7.3 Raising Animals for Human Consumption from an Efficient Perspective

Taking into consideration the situation in developed nations, there has been a shift in animal husbandry from grass-fed to grain-fed livestock. This means that animals that were once dependent on grass are now being fed grain crops. This shift highlights a comparison between the amount of grains required for animal consumption by humans (animal protein) and that required for human consumption (plant protein). For chicken, the ratio is 2.3, and for beef, it is 13. This means that grains consumed for animal protein are 2.3 times and 13 times more than those for plant protein, respectively.

---

and sustainable diets united against hunger; 3–5 Nov 2010; Rome, Italy. Rome, Italy: Food and Agriculture Organization, 2012. https://www.fao.org/4/i3004e/i3004e.pdf

Additionally, this shift places an extra burden on fossil fuel energy, with researchers claiming that 11 times more fossil fuel energy is required to produce animal protein than plant protein. This raises an important question: Is raising animals for human consumption efficient? The energy required for the cultivation of animal protein is significantly higher (4 times more for chicken and 40 times more for beef) than for plant protein. Moreover, the utilization of land and water for producing soy as a source of protein is much less compared to that for animal protein. Thus, it is difficult to argue that raising animals for human consumption is efficient.

The emission of greenhouse gases is a central factor in environmental degradation. There are various stages of food production, including farming, production, processing, packaging, transportation, marketing, sales, and purchasing. All these stages contribute to greenhouse gas emissions. However, if one has to identify which stage is the highest contributor, the answer would be farming. The farming of livestock emits nitrous oxide and methane, accounting for 80% of greenhouse gas emissions. When comparing animal protein to plant protein such as soy, the emissions are significantly higher for animal protein. Emissions of acidifying substances are 7 times higher, pesticides 6 times higher, and metals 100 times higher for animal protein compared to plant protein (soy-9).

As the population increases, so does the demand for food, particularly animal-based food. Meat consumption has exponentially increased by 63% worldwide and over 300% in developing nations since 1963. By 2050, the world population is expected to reach 9 billion. A plant-based diet requires fewer natural resources and is less harmful to the ecological balance. This chapter argues that there is a dire need for a dietary shift, specifically a significant reduction in the consumption of animal-based foods, as many people prefer meat and dairy products. Historical evidence shows that the shift to a plant-based diet has occurred either out of necessity or by choice, often advocated for religious, ethical, or philosophical reasons. It is only in the last 150 years that advocacy efforts have taken on a scientific approach.

Returning to a plant-based diet on a global scale is expected to face political challenges, with optimization of food supply, environment, health, and social justice being one of the ways to address this issue. This chapter focuses on the aspect of environmental sustainability and does not address the issues of nutritional, cultural acceptability, and economic accessibility.[18]

According to Invest India's report on "Cultivated Meat: Key Opportunities and Recommendations in India," the global demand for protein is projected to double by 2050. This increase is driven by changing food consumption patterns due to rising income levels and growing urbanization. A significant portion of this demand will come from emerging markets like India, where the demand for poultry meat alone is expected to increase by 850 percent by 2040. The report provides an overview of the meat alternative known as "cultivated meat," which includes plant-based, cultivated, and fermentation-derived proteins.

[18] Sustainability of plant-based diets: back to the future https://www.sciencedirect.com/science/article/pii/S0002916523048992

Globally, policy trends reflect a growing focus on cultivated meat. In 2020, Japan declared smart proteins an "important sector" and formed the Japan Association for Cellular Agriculture (JACA). JACA offers a collaborative platform for industry, government, and academia to develop guidelines and recommendations for laws applicable to cultivated meat, egg, and dairy products, facilitating their commercialization in Japan. Similarly, the UK has recognized cultivated meat and plant-based meat alternatives as significant advancements to address food shortages and climate change.

7.4 Overview of Policy Regulations Promoting Sustainable Meat Substitutes Across Various Countries[19]

Food production accounts for nearly 35% of total greenhouse gas emissions, with 57% of these emissions attributed to animal-based food production, including livestock feed. This section outlines various policy measures taken by countries to establish a regulatory environment that promotes sustainable meat substitutes.

7.4.1 The USA

In 2018, the US Department of Agriculture (USDA) and the US Food and Drug Administration (FDA) announced a regulatory framework for cultivated meat. The FDA oversees cell collections, cell banks, and all cultivation inputs, while the USDA oversees further processing and labeling of food products.

7.4.2 European Union

The EU supports and promotes smart proteins through its regulatory framework, stating, "New technologies and innovations in food production should be encouraged as they could reduce the environmental impact of food production, enhance food security, and bring benefits to consumers, as long as a high level of consumer protection is ensured." This is categorized under Regulation (EU) 2015/2283 on novel foods, which explicitly refers to food produced from "cell culture or tissue culture derived from animals."

[19] Invest India, "Cultivated Meat: Key Opportunities And Recommendations in India", Report on the technical and market landscape of cultivated meat with key policy recommendations for India, 2023

7.4.3 Singapore

The "Requirements for Safety Assessment of Novel Foods" is the current regulatory framework in Singapore, covering the manufacture and sale of cultivated meat products. In November 2020, Singapore became the first nation to greenlight the sale of a cultivated meat product.

7.4.4 India

The Food Safety and Standards (Approval of Non-Specified Foods and Food Ingredients) Regulations 2017, regulated by the Food Safety and Standards Authority of India (FSSAI), lays down the pre-market authorization for novel food products. This includes food ingredients using novel technology, new additives, or new processing methods, including enzymes not previously regulated, and food ingredients consisting of isolated microorganisms, bacteria, yeast, fungi, or algae. Although the definition is broad and does not specifically mention plant-based meat alternatives, there is a section requiring information on whether the food product is from an animal source or not. The ASSOCHAM Report highlights the burgeoning market for plant-based alternatives, including plant-based meat and vegan products. It identifies this sector as a rapidly growing industry with significant potential, describing it as a "sunrise sector" full of opportunities. According to the report, India's market for plant-based meat is projected to grow exponentially, increasing from the current estimate of $30–40 million to an impressive $500 million.[20]

7.5 Policy Recommendations to Boost the Production of Plant-Based Meat Alternatives and Vegan Products

1. Financial Incentives: The government provides financial support to companies producing and marketing vegan products and plant-based meat alternatives. This could be achieved through tax breaks, grants, and loans, offering a financial cushion to these companies and encouraging their growth.
2. Research and Development: The urgent need for the government to fund research and development programs aimed at enhancing the taste, texture, and nutritional profile of vegan products. This investment in R&D is crucial for making plant-based alternatives more appealing to consumers.

[20]The Associated Chambers of Commerce and Industry of India (ASSOCHAM) Report, "The Rising Plant Based Sector in India: Insights and Opportunities", February 2023, https://www.assocham.org/uploads/files/The%20Rising%20Plant%20Based%20Report%20final%20for%20approval.pdf

3. Education and Awareness: To increase the adoption of vegan products and plant-based meat alternatives, the government should focus on educational campaigns. These campaigns would raise awareness about the benefits of these products, helping to shift consumer preferences toward plant-based options.
4. Regulatory Support: Simplifying the regulatory landscape is another key recommendation. The government should implement policies that facilitate the marketing and sale of vegan products. Ensuring accurate labeling and removing barriers to the sale of these products will make it easier for consumers to choose plant-based options.
5. Infrastructure Investment: Finally, significant investment in infrastructure is needed to support the production and distribution of vegan products and plant-based meat alternatives. This includes developing cold storage facilities, transportation networks, and processing plants. Such infrastructural investments are essential for building a robust supply chain and ensuring the efficient distribution of plant-based products.

7.6 Consumer Trends for Plant-Based Meat Substitutes

The critical factors influencing consumers' food choices are multifaceted. For most consumers, taste is paramount, whereas for those with lower incomes, price becomes the most significant factor. Considering this, the introduction of taxes to increase the price of meat could be a viable strategy to reduce meat consumption. However, the research paper notes that there is no existing study that has observed the impact of such taxes on meat consumption in a real-world setting.

One proposed government intervention involves subsidizing healthier, vegan, or plant-based meat alternatives to make them 20% to 50% cheaper compared to less healthy products. This price reduction aims to create a substitution effect, encouraging the consumption of plant-based meat alternatives over traditional meat. Despite this, it is unlikely that such high taxes on meat would be readily accepted by consumers or producers. Policymakers and politicians may also be reluctant to introduce such taxes, predicting that any acceptable tax increase would have a minimal effect on reducing meat consumption. Thus, the focus remains on promoting plant-based meat alternatives for healthier lifestyles and environmental benefits.

Similar to how nutritional information is provided on food packaging, sustainability information could also be included. While a few studies have conclusively shown that labeling impacts consumer buying behavior, the report indicates that stronger interventions could influence behavior, even if only in the short term.

Another recommendation from the research paper is labeling climate-friendly dishes in canteens and providing information about their nutritional and climate impacts. This approach is likely to increase the sales of such products significantly. This conclusion supports the idea that more policy interventions from government or regulatory authorities could effectively promote plant-based meat alternatives.

The research paper highlights that meat consumption is associated with negative environmental and animal welfare consequences. Scientists often argue about the urgent need to reduce meat consumption. Enhancing the taste, texture, and nutritional value of meat substitutes can give plant-based meat alternatives a competitive edge, making them more acceptable to a broader range of consumers. However, it is not safe to assume that increased consumption of plant-based meat alternatives will automatically lead to reduced meat consumption.

The global demand for protein is set to double by 2050 due to rising income levels and increasing urbanization, with significant implications for emerging markets like India. The Invest India report on "Cultivated Meat: Key Opportunities and Recommendations in India" highlights that the demand for poultry meat in India is expected to surge by 850% by 2040. To address this demand, the report emphasizes the potential of cultivated meat, including plant-based, cultivated, and fermentation-derived proteins.

Internationally, countries are proactively developing supportive policies for the cultivated meat sector. Japan has designated smart proteins as an important sector and established the Japan Association for Cellular Agriculture to develop commercialization guidelines. The UK has recognized cultivated meat and plant-based alternatives as crucial for addressing food shortages and climate change. In the USA, the USDA and FDA have created a comprehensive regulatory framework, dividing responsibilities between cell cultivation and food processing/labeling. The European Union promotes smart proteins through Regulation (EU) 2015/2283 on novel foods, and Singapore has implemented a framework for the safety assessment of novel foods, becoming the first nation to approve the sale of cultivated meat.

In India, the Food Safety and Standards Authority of India (FSSAI) regulates novel food products, including plant-based meat alternatives, under the Food Safety and Standards (Approval of Non-Specified Foods and Food Ingredients) Regulations 2017. The ASSOCHAM report highlights the potential for India's plant-based meat market to grow from $30–40 million to $500 million, identifying it as a "sunrise sector" with immense opportunities.

To capitalize on this potential, several key policy recommendations are made:

1. Building R&D and Manufacturing Capacity: Central and state governments should support startups and companies through infrastructure development, financial support, and incentives.
2. Cross-Country Collaboration: Partnerships with global entities can enhance food technology and innovation, supporting India's sustainable development goals (SDGs).
3. Building Regulatory Clarity and Preparedness: Clear regulations specific to plant-based meat alternatives are essential to attract large-scale investments.

The focus should also be on policy considerations for regulating subsidies on meat alternatives. Introducing taxes on conventional meat and subsidizing plant-based alternatives could influence consumer demand. However, labeling and educational campaigns might be more readily accepted and effective in promoting plant-based meat alternatives.

The urgent need to reduce meat consumption due to its negative environmental and animal welfare impacts is clear. Enhancing the taste, texture, and nutritional value of meat substitutes is crucial for broader consumer acceptance. By implementing these recommendations, India can significantly boost the production and consumption of plant-based foods, positioning itself as a leader in the global market for sustainable meat substitutes.

Chapter 8
Overview of Current Regulations and Labeling Requirements for Meat Substitutes

Laws regulating meat alternatives are changing significantly, as the market for plant-based proteins expands. There has been a lot of innovation in the plant-based industry, especially in terms of increasing public acceptability and knowledge. Companies are spending money on improved product designs and packaging that appeal to local and cultural tastes.

While early adopters in India are somewhat knowledgeable about plant-based products, there is still a significant knowledge gap regarding the category and the many names used by producers. This problem is most noticeable in the plant-based dairy industry, where consumers need clarification about how to use products due to the lack of traditional words like milk, cheese, and butter. Prominent labeling with terms such as "plant-based" and "vegan" helps differentiate these products from traditional meat and dairy. Additionally, consumers often rely on vegetarian and nonvegetarian logos to distinguish plant-based items from those derived from animals.

Indicating the protein source helps consumers accurately identify and visualize the sensory attributes such as appearance, aroma, taste, texture, and aftertaste; with this, consumers can understand. Visual cues like packaging shape, color, and imagery convey category-relevant information to consumers. Images on the packaging can help consumers identify the category and understand the product's look and feel when traditional dairy terms are not used.

Taste and texture are vital for consumers to appreciate plant-based products like meat or cheese. Consumers gauge these attributes through multiple cues, including visuals, listed ingredients, and descriptions on the packaging. Hence, it is recommended to use appealing images along with indulgent descriptions. Health claims are particularly significant for consumers of plant-based milk alternatives, while taste claims resonate more with those interested in plant-based meat alternatives. The high protein content is another essential claim within the plant-based category.

S. Ray, S. Suri, *The Sustainable Plate: Unveiling the Science of Meat Substitutes and Impact on Global Health*, https://doi.org/10.1007/978-3-031-94604-2_8

Claims undermining conventional animal-derived products should be avoided because the target audience frequently eats plant-based and animal-derived items [1].

There are no particular rules or regulations pertaining to plant-based dairy or meat products in India. These products must instead follow the basic labeling guidelines that apply to all food items. Plant-based meat substitutes must therefore be clearly marked and packed with information regarding their plant-based origin and attributes. When using terms like "plant-based" or "soy-meat" in the product name, labels should clearly state that the product is free of animal meat. Additionally, pictorial representations can be used on packaging to demonstrate the absence of animal meat.

On July 29, 2020, the FSSAI issued a draft notification proposing a ban on "dairy terms" or similar terms for "dairy analogue" products, including plant-based almond milk, oat milk, yogurt, cheese, and butter. The Good Food Institute and Ikigai Law argued that this prohibition should not apply to plant-based dairy substitutes. If implemented, the warning might have a big impact on the plant-based dairy industry since it forbids the use of dairy-related terms in product names. Therefore, there was still regulatory ambiguity until the FSSAI established clear criteria for plant-based meat replacements [2].

In 2022, the Food Safety and Standards Authority of India (FSSAI) published its initial vegan food regulations. These rules specify the requirements that plant-based products must meet as well as the labeling criteria that must be adhered to. The new rules define vegan cuisine as having no animal products of any type and no animal components used in its production or preparation. Only with the FSSAI's approval can vegan products be created through animal testing. Businesses may now more confidently market their products as truly vegan, appealing to consumers who are concerned about their health and morals, thanks to this clarity [3].

Companies are requesting more stringent regulations from the Food Safety and Standards Authority of India (FSSAI). Uncertain legislation has led to varying interpretations, particularly with relation to the proposed prohibition on the use of dairy-related terminology for plant-based products. More specific regulations, according to stakeholders, would encourage industry growth and help prevent future legal problems.

Singapore has implemented "The Requirements for Safety Assessment of Novel Foods" to regulate cultivated meat. This framework describes cultivated meat as "meat developed from animal cell culture," involving the growth of selected cell lines or stem cells in a bioreactor with appropriate growth media, which may be assembled on a scaffold to create products resembling muscle meat. Manufacturers, importers, or sellers of cultivated meat must submit a safety assessment to the Singapore Food Agency (SFA) for approval. The SFA prioritizes product safety over the production process and does not charge fees for evaluating applications, typically completing the assessment within 3 to 6 months. Products must be labeled with terms like "cultured" to clarify their origin for consumers. Confidential business information provided to the SFA is protected and not disclosed without consent.

Entrepreneurs in the cultivated meat sector are encouraged to engage with the SFA early in the product development phase to understand the safety requirements.

The extensive regulatory system in Singapore outlines particular standards for innovative foods, such as grown meat, and acknowledges subcategories within these goods. With information on cell-line origins and production scaffolding, this clarity guarantees a comprehensive safety evaluation.

As they create regulations for new food categories, authorities in India and other nations can use this framework as a useful guide, which offers a thorough and comprehensive method for guaranteeing product safety and clarity [4].

Food Standards Australia New Zealand (FSANZ) controls cultured meat in Australia and New Zealand, setting it apart from plant-based substitutes. According to FSANZ's Food Standards Code, farmed meat produced using animal cell culture technology is considered a novel food. To recreate traditional meat products, like steak or minced meat, without recreating the entire animal, FSANZ defines cultured meat as meat made from animal cells utilizing sophisticated biotechnological and synthetic procedures.

Before cultivated meat can be marketed, it must pass a health and safety assessment. This includes evaluating potential health risks, the product's composition and structure, the methods used in its production, and labeling details. The product can be sold only after this assessment and inclusion in the "Permitted Novel Foods" register. FSANZ may impose specific conditions, such as detailed preparation instructions or compositional standards. The length of the premarket approval process is currently not specified by FSANZ, which would aid businesses in their preparation [5].

The Committee on Agriculture & Rural Development of the European Parliament suggested in April 2019 that terminology like "burger," "steak," and "sausage" be limited to goods made entirely of animals, excluding plant-based substitutes. This change's proponents contend that it is essential to give customers precise and understandable information about the types of goods they are buying [6].

A British nonprofit group called The Vegan Society opposed the change, claiming it would breach EU accurate labeling requirements and violate customers' access to knowledge. In a vote on October 23, 2020, the entire EU Parliament rejected the amendment despite receiving approval from the Committee. For the brands that are currently available on the market in EU nations and use terminology like sausages and burgers for their vegan or vegetarian products, the restriction would have resulted in expenses and other negative effects. The tension between some foods' legal, cultural, and conventional definitions would have increased with the prohibition [7].

In order to prevent "misleading claims" of certain meat substitutes, France published updated recommendations on September 4, 2023, prohibiting the use of meat names for domestically produced plant-based foods. "It is an issue of transparency and fairness which responds to a legitimate expectation of consumers and producers," said France's Ministry of Agriculture and Food Sovereignty in a press statement [8].

The US Food and Drug Administration (FDA) and the US Department of Agriculture (USDA) are the two organizations in charge of regulating food labeling in the United States. While the USDA governs meat and poultry products, including many farmed meats, the FDA regulates the majority of food items, including plant-based proteins. Businesses must adhere to the labeling regulations set forth by both organizations.

The Federal Food, Drug, and Cosmetic Act (FDCA) gives the FDA authority to regulate food labeling. According to section 403(g) (1) of the FDCA, a food is considered misbranded if it does not meet the FDA's identity standards. According to section 403(i) (1), labels for goods without a standard of identification, like alternative proteins, must include a "common or usual name" that unambiguously identifies the product. This name should not be confusingly similar to other food names and should be consistent for similar goods. If there is no widespread name, the label can employ a term that describes something or a popular, imaginative name.

Food labels are not preapproved by the FDA, but it does expect adherence to rules that include truthful depiction and clarity to prevent consumer fraud. Legal action against manufacturers or enforcement measures like warning letters may result from any deceptive labeling [9].

Under the Federal Meat Inspection Act and the Poultry Products Inspection Act, the USDA, through its Food Safety Inspection Service (FSIS), controls the labeling of meat and poultry. Additionally, labels for these products must follow specific identity criteria and be truthful rather than deceptive. Before being marketed, labels—including those with unique claims or statements—are examined and approved by FSIS. Products made from farmed beef must pass a premarket evaluation process and have thorough paperwork to back up any unique claims.

Although there are currently no explicit standards in place, the FDA and USDA are trying to develop unified labeling principles for cell-cultured products. FSIS is considering a petition to restrict the term "meat" to traditionally harvested animal tissue and has asked the public for feedback on identifying meat and poultry made using cell culture technology [10].

This terminology issue is especially important as authorities debate whether labels like "meat" should only refer to animal-derived products or if they can also refer to cell-cultured alternatives. Due to legislative ambiguity, determining the extent of labeling for novel food categories, such as produced meat, is still extremely difficult.

Plant-based products are governed in the United States by several certification organizations, such as the American Vegetarian Association, Vegan Action, BeVeg, and the Plant Based Foods Association. With the National Sanitation Foundation International acting as the certifying body, the Plant Based Foods Association introduced the first "Certified Plant Based Seal" program in the United States in 2018. The seal is intended to make it easier for consumers to identify plant-based products that follow stringent standards for plant-based component and production procedures.

Because of the FDA's lax labeling regulations, terms like "milk," "cheese," and "meat" can be used alongside qualifiers like "vegan" or "plant-based." However, since any potential misunderstanding or dishonest activity may result in penalties, customers must not be deceived by these indications. It is expected that the substitute foods' nature will be clear from the labels. Words like "almond milk" or "vegan cheese," for instance, are acceptable as long as they are appropriately qualified and let consumers know that the product is made from plants [11].

The Plant Based Foods Association has developed voluntary labeling guidelines to promote consistency in the labeling of plant-based products that adhere to state and federal regulations. They created rules for plant-based milk in 2018, guidelines for plant-based yogurt in 2020, and guidelines for meat substitutes in 2019. These guidelines are meant to ensure that plant-based goods are labeled in a way that reduces customer confusion and promotes transparency regarding the ingredients and production methods [12].

Research on cultured meat in South Korea is still in its early stages and lags behind global advancements by about 4 to 5 years. Despite this, a growing number of companies are actively developing cultured meat, signaling potential market growth as major firms and start-ups like CJ CheilJedang, Dae Sang, SpaceF, Cell Meat, DaNAgreen, and SeaWith invest in this area. It has been proposed that cultured meat should be classified as a new food category and recognized as a distinct industry separate from traditional livestock. Additionally, there is a suggestion to introduce new terminology such as "cell-derived protein foods" rather than labeling these products as "meat" to better reflect their unique nature [12].

The laws governing grown meat in South Korea have been a major obstacle despite recent improvements. Cultivated meat had not been approved for commercial sale until recently, and it was only used for research and development. By creating the Gyeongbuk Cell-Cultivated Foods Regulatory-Free Special Zone (RFSZ), the government has taken action to address this issue and promote innovation in the industry while reducing legal barriers. However, businesses trying to approve their items still face difficulties because of the current laws [13].

Currently, it is not economically feasible to produce cultured meat in large quantities due to the need for costly growth media and bioreactor systems. To cut these expenses, businesses like SeaWith are investigating cutting-edge techniques like employing microalgae as a growth medium [14].

In Japan, plant-derived meat falls under the category of "other processed foods" as per the Classification of Processed Foods, with a strong emphasis on accurate raw material labeling. On February 24, 2022, the Japanese Agricultural Standards (JAS 0019) introduced "Textured Soy Protein Products" to the list maintained by the Minister of Agriculture, Forestry and Fisheries. This addition established certification technology standards and inspection methods, setting precise requirements for labeling "soy meat products" and "prepared soy meat products."

Under the Food Labelling Standards in Japan, retail processed foods must be labeled with "general names," which cannot include ingredients not present in the product. Consequently, foods made from soybeans cannot be labeled as "alternative meat," leading companies to use terms like "soybean-based processed foods" instead. Labels must also clearly state that the product is not actual meat, with this information prominently displayed on the packaging. Common terms like "meat" and "egg" are prohibited in the raw material names of plant-based products. However, terms such as "soy-based meat" or "oat milk" are permitted as long as the label explicitly states that the product is not animal meat, such as by noting "No meat" or "soybeans used as raw material." There are also provisions allowing claims like "100% plant-based," provided all ingredients are indeed derived from plants [12].

Similarly, the Dietary Guidelines for Chinese in China strongly emphasize the necessity of clearly labeling foods, especially those that are plant-based or alternative. The National Health Commission and other pertinent authorities created the guidelines as part of a comprehensive framework to raise knowledge of nutrition and public health. In order to guarantee that people are aware of what they are eating, products bearing labels such as "plant-based" must also comply with stringent laws governing ingredient disclosure. This is a component of a more significant effort to lower chronic diseases linked to poor nutrition and enhance general public health in China [15].

In Australia, Vegan Australia manages the certification process for meat replacement products. Companies that meet the criteria can use the vegan certification logo for a year, with the option to renew it annually. This certification covers both entirely plant-based products and those that include plant-based ingredients mixed with animal-derived components, as specified by the Food Standards Regulations and Food Standards Code (Section 1.1.1—13(4)). Similarly, New Zealand recognizes plant-based alternatives, while Singapore has made history by approving the world's first cell-cultured food.

Singapore's regulations for new food safety assessments mandate the use of descriptors like "mock," "cultured," "imitation," or "plant-based" to ensure clarity when the product is purely plant-based. The SFA stresses that alternative protein products ought to be labeled with precise wording. This rule is intended to avoid customer confusion and guarantee that goods labeled as plant-based or comparable appropriately represent the ingredients they contain [16]. Plant-based foods and other novel foods must undergo a premarket safety evaluation procedure. Before items are put on the market, this procedure assesses possible threats to food safety, such as toxicity and dietary exposure [17].

In South Korea, discussions are ongoing within the livestock industry regarding the appropriate terminology and labeling for these emerging food categories [12].

International manufacturers of plant-based meat alternatives should note that European countries are adopting stricter labeling regulations. France has banned terms like "steak" or "sausage" for products not containing animal meat, with fines up to €300,000 for noncompliance.

In the same way, the Agriculture Committee of the European Union Parliament has proposed legislation to prohibit the use of traditional meat terminology for plant-based products throughout the EU. The Council of Ministers and the European Parliament have not approved this law. If approved, it would prohibit companies from using phrases typically associated with animal-based products to identify plant-based meat substitutes, further complicating the regulatory landscape for global corporations [18].

At present, there is instability in the regulatory framework around the labeling of plant-based meat products. State laws have become inconsistent due to the lack of federal regulation, and these provisions may be challenged in ongoing federal litigation. International restrictions can make things more complicated. Manufacturers must continue to monitor and adjust proactively to maintain compliance with the quickly evolving state and, maybe, federal requirements [18].

8.1 Addressing the Challenges and Opportunities for Policy Development in Promoting Sustainable Meat Substitutes

There are numerous strategies to reduce the average amount of meat consumed, and one of them is by consuming more plant-based meat alternatives (PBMAs). Because of this, hundreds of products with sensory qualities (taste, texture, appearance, and smell) comparable to those of their animal counterparts have been introduced to the market in recent years [19].

The development and marketing of sustainable meat alternatives offer both enormous challenges and exciting prospects for the formulation of public policy. The environmental, moral, and health concerns related to conventional meat production are being addressed by these substitutes, which include plant-based proteins, cultured meat, and insect-based products. An obstacle that often arises in the development of plant-based meat alternatives is the replication of the sensory properties of the cut meat, such as the design, texture, smell, and taste. This obstacle is especially acute for regions with deep-seated traditions and habits that revolve around the use of meat in the preparation of meals. These sensory attributes are fundamental in the filling of the target markets. However, the market for meat substitutes is fraught with difficulties. The industry went through turmoil in 2023 owing to stiff competition from beef products and an absence of private funding. Production efficiency is a significant concern for them as the market has become overly competitive and oversaturated [19]. In order to achieve higher mass consumption of plant-based meats, real meat-eating experiences should be replicated even in more aggressive cooking conditions. However, achieving a similar bite's "attack" is still infrequent and usually calls for flavor enhancers. These solutions could help resolve the negative impression over the perception of PBMAs as unnatural, highly processed, and so on, rather than actual meat. These additives can contribute to a perception that PBMAs are highly processed or unnatural, which may deter consumers.

Additionally, the general public may find PBMAs less appealing if they are inexperienced with them, particularly in how to cook and prepare them. One major obstacle to acceptance is a lack of familiarity with the product. In order to create PBMAs that closely mimic traditional meat in terms of sensory experience and usability, sensory and consumer studies must be fully integrated. Direct consumer experience with PBMAs is crucial to overcoming these barriers and expanding their market reach [20].

Due to their more significant dietary needs, plant-based businesses are moving from vegans and flexitarians to omnivores. These omnivores are meat eaters open to trying plant-based foods while reducing their meat intake. Taste is important because items must appeal to meat eaters in order to maintain their interest in the category [19].

According to producers of plant-based foods, two other crucial elements must be met. One is enhancing brand awareness so that consumers can better remember the brand if a product has a rancid flavor. Second, more has to be done to promote plant-based meats' health and environmental advantages because many products contain soy and are fortified to improve nutrition [19].

As the supply of cultured and plant-based meats would be expected, such advancement could have drastic benefits and challenges concerning the environment, society, and economy. In relation to these alternative meat products, the overall environmental impact is expected to be less than that of animal meat, including less GHG emissions and land and water utilization. On the contrary, some research points out that the energy use of cultured meat may be higher than that of traditional animal meat. These goods may also significantly lessen the problems with animal welfare related to the production of meat. They would lessen the usage of antibiotics and the spread of infections through food, among other public health concerns.

Additionally, there may be economical repercussions from the growing popularity of plant-based and cultured meat, particularly in regions where animal agriculture plays a major role in the rural economy. The traditions and means of subsistence of rural populations may be disrupted by these industries. However, they may also create new jobs and economic opportunities associated with the developing alternative meat market and its supply networks. The social and economic impacts on rural communities, especially those tied to animal and crop agriculture, have been less explored but could involve challenges and new prospects [21].

In the European Union, policymakers actively support developing and promoting alternative meat substitutes and hybrid products. The Farm to Fork Strategy is an example of how this is accomplished by sponsoring research initiatives promoting more sustainable protein sources. However, understanding how consumers view these items continues to be a major obstacle for the food sector. Creating meat substitutes that appeal to people requires a thorough strategy that incorporates consumer and sensory research [20].

Policymakers in European countries should prioritize research to understand better the motivations and barriers to consuming meat substitutes in different regions. They can then use these insights to craft strategies that reduce meat consumption and boost the adoption of meat substitutes. This could include promotional activities, nudges, and support for meat substitute businesses. Research indicates that informing consumers about meat substitutes can increase acceptance, so a large-scale marketing campaign highlighting their benefits—such as health advantages—could be effective. Regular meat substitutes at canteens and cafeterias, along with special offers or reduced costs, may also assist overcome long-standing opposition and encourage more people to make the move [22].

Price parity with conventional meat is a significant challenge for the mass adoption of plant-based meat. According to FMI research, nearly a third of consumers cite cost as a primary reason for discontinuing their plant-based meat or dairy purchase. The premium pricing of plant-based foods limits their appeal to a broader audience, especially as consumers become more price-conscious in the current economic climate [23].

Nonetheless, a study published by ProVeg Netherlands suggests that the difference in cost between traditional and plant-based meats is shrinking. In the last 5 years, there have been plant-based alternatives to over half of animal-derived products that are either comparable in price (33%) or less expensive (20%) than conventional alternatives. Plant-based goods can save customers up to 20% at certain Dutch supermarkets when compared to their animal-based counterparts [24].

Notwithstanding economic difficulties, the market for plant-based meat has expanded significantly. For instance, Australia saw a 47% increase in sales and a 28% increase in per capita consumption between 2020 and 2023. Although price is still an essential factor, this trend shows that plant-based products are growing in popularity [25].

In line with the SDGs and the Paris Climate Agreement, the World Economic Forum research highlights the significance of a specific "Transition Decade" from 2020 to 2030 in order to move our global protein system toward sustainability. This transition from traditional meat to alternative proteins will unfold differently worldwide, and relying solely on market forces or isolated innovations will not be enough to provide affordable, healthy, and sustainable protein by 2030.

By 2050, it is anticipated that the rising demand for meat will have increased by at least 50%, posing severe environmental problems such as increased pollution and resource depletion. Compared to traditional meat production, alternative proteins—derived from plants, fermentation, or cultured animal cells—offer a more sustainable option since they use much less water and land. They are essential for reducing the effects of climate change since they also produce fewer greenhouse gas emissions (GHGs). A significant investment in research and infrastructure is required to make the transition to alternative protein manufacture. Public investments have the potential to create a substantial amount of economic value and job growth in the industry. Therefore, politicians are encouraged to support this change. By 2050, for example, switching to alternative proteins is predicted to provide up to 83 million jobs worldwide [26].

The deeply personal and cultural significance of meat makes broad policy changes challenging. In many cultures, meat is a symbol of tradition and identity and a staple food. For instance, beef is a significant part of Argentina's pride and culinary heritage; thus, any attempt to reduce cattle consumption would be contentious and politically sensitive. The same applies in Texas, where beef and barbecue are central to local culture and social gatherings, leading to considerable opposition to legislation limiting their use or production [27].

Argentina's beef industry plays a significant role in the nation's economy by generating jobs and assisting rural communities. In Kenya, raising cattle traditionally is crucial to livelihoods and food security. Reducing beef consumption in Argentina or Texas or replacing livestock farms with lab-grown meat in Kenya can meet substantial resistance due to meat's cultural and economic importance. According to research, awareness of "artificial meat" is rising, but people are still not very keen to taste it, especially older people who might think it is inferior or unnatural. The idea that lab-grown beef is "unnatural" may cause customers who value conventional food sources to become psychologically resistant [28]. As a result, effective methods require appealing narratives that match people's values and fulfill their current needs.

Different regions require tailored approaches. For example, China's investment in cultured meat is a component of a larger plan to improve food security and lessen its impact on the environment. To allay worries about food safety, the narratives emphasize controlled manufacturing facilities. Before introducing more reasonably priced alternatives, early adopters in the United States may be drawn in if cultured beef is first positioned as a luxury good, similar to expensive electric vehicles.

In order to speed up this transformation, it is critical to create integrated public-private narratives and encourage intergovernmental engagement. Creating integrated public-private narratives and encouraging intergovernmental cooperation is critical to accelerating this change. It is critical to combat counter-narratives and ensure that the messaging is clear and supported by research. There is room for innovation in the conventional meat industry, even though traditional and alternative proteins will continue to coexist for some time. A unified "Transition Decade" strategy will be vital to advancing technology, financing, and policy solutions, making sustainable protein options both appealing and widely accessible [29].

According to a recent Good Food Institute research, governments throughout the world are investing almost $1 billion in public partnerships, indicating that alternative proteins are becoming more and more important in national food security plans. Public investment for alternative proteins has totaled $1.67 billion since 2021, with an expected $523 million allocated globally in 2023. This funding helps the alternative protein sector with infrastructure development, commercialization, and research. Significant increases in funding by nations like the United Kingdom, Germany, and Canada show that alternative proteins are increasingly being acknowledged as an essential part of national plans [30].

Governments have greatly benefited from the Good Food Institute's (GFI) guidance in developing national policies for alternative proteins. Their three-step approach, tailored to each country's unique strengths and objectives, focuses on motivation, vision, and activity. This strategy seeks to improve food security and sustainability by bringing public stakeholders into line with current policy [30].

Governments around the world are progressively supporting alternative proteins in reaction to the volatility of food prices in 2022, which highlighted the effect of geopolitical tensions on food security. Developing this sector is a priority for nations like Israel and Singapore in order to increase self-sufficiency and build a more robust agricultural sector that can endure obstacles like export restrictions, war, and climate change.

Investing in alternative proteins not only supports food sovereignty goals but also offers significant economic potential. Continued investment in cellular agriculture—the technology behind lab-grown meat—could generate $1.1 trillion in global economic activity and create 9.8 million jobs by 2050. For comparison, the global meat industry was valued at $897.5 billion in 2021.

The GFI report highlights that significantly more investment is required to scale alternative proteins effectively—around $10.1 billion in total. Both plant-based and cultured meat sectors depend on government funding to transition from the lab to the table in substantial quantities.

To make cellular agriculture widely accessible, securing regulatory approval is just the first step. The development of the plant-based sector is primarily dependent on government support. It promotes novel techniques that improve the flavor and cost of these products, allowing them to compete with traditional meat. It also helps the development of crucial infrastructure, such as the sophisticated bioreactor factories required to produce cultured meat products. In addition, public investment boosts market confidence by notifying private investors about the demand for alternative proteins. This encourages private investment and accelerates the sector's growth [31].

Singapore and Israel lead the charge in alternative protein investments and regulatory advancements. Temasek, Singapore's state owned innovation investment fund, ranks as a top global investor in this sector. Singapore has also formed partnerships with international universities, such as the University of Bath in the United Kingdom, and has set regulatory precedents, with the Singapore Food Agency becoming the first to approve the sale of cultivated chicken from Good Meat in 2020. Israel has established the largest government-backed consortium for cultivated meat, funded by $18 million from the Israel Innovation Authority in 2022. This consortium includes 10 academic labs and 14 companies, such as TNUVA, Israel's largest food manufacturer. Israel also supports open-access domestic and international facilities, using alternative protein development as a diplomatic tool.

Meanwhile, the Netherlands has funded a cellular agriculture ecosystem, which includes research, development, commercialization, and workforce transition programs. China has held significant events like the China Cellular Agriculture Forum and a panel on cellular agriculture at the China International Food Safety and Quality Conference.

In contrast, the United States has been slower in supporting cultivated meat development. Although President Biden has publicly endorsed it, Congress has only provided $6 million to the USDA's Agricultural Research Service for alternative protein research. The National Institute of Food and Agriculture (NIFA) allocated $10 million in 2021 to Tufts University's Center for Cellular Agriculture, the first US lab dedicated to this field. However, this funding is modest compared to Israel's $18 million investment, despite the United States having the world's largest GDP [32].

India confronts many obstacles in promoting sustainable beef alternatives, such as a lack of political agreement, bureaucratic roadblocks, cultural aversion to changing eating patterns, and a lag between policy creation and action. However, there are many opportunities to develop these initiatives. Growing consumer awareness and demand for plant-based diets, international collaboration, public-private partnerships, and technological advancements in food production can all significantly advance sustainable alternatives. Such programs can also be made more effective by being integrated into larger agricultural and environmental agendas. By addressing these concerns and capitalizing on existing opportunities, India can build a more sustainable food system with sustainable beef alternatives [33].

8.2 Stakeholder Engagement and Collaboration for the Future of Meat Substitutes

Effective stakeholder involvement and cooperation across various industries, including agriculture, food technology, consumer advocacy, and policy-making, are becoming increasingly crucial for the future of meat substitutes. The responsibilities of various stakeholders become increasingly important in determining the landscape of meat substitutes as the demand for sustainable protein sources increases. Businesses must be responsible for providing high-quality, affordable, and popular plant-based foods, a fast-growing industry with consumers' desire for less meat consumption. The research community can play a crucial role by encouraging a diverse range of scientists across various disciplines and career stages to enter the alternative protein field. Governments worldwide should also invest the required funds to support the growth of R&D and manufacturing scale-up, give incentives for large-scale manufacturing, and set up competitive market conditions in which plant-based meats are equal to traditional ones in terms of taste, price, and convenience. These actions can help to tackle the significant problems with plant-based meat, encourage more research, identify emerging sectors, and ensure that the benefits of plant-based meat are communicated across different populations and societies [23].

Overview of plant-based meat's sales performance in US retail is as follows: (1). Sales Declines: The category is experiencing declines due to a slight drop in consumer purchases, particularly in the refrigerated subcategory, as inflation tightens budgets and makes plant-based price premiums more noticeable. (2). Consumer Loyalty: Despite higher prices, households that continue to buy plant-based meat purchase consistently compared to the previous year, indicating strong loyalty among existing consumers. (3). Growth in Emerging Subcategories: Growth within the category is driven by new product launches in emerging subcategories such as plant-based chicken, seafood, and jerky [23].

Despite recent declines in sales, consumer interest in reducing meat consumption and increasing plant-based food intake remains strong. Research by Credit Suisse shows that 66% of consumers aged 16–40 across 10 countries plan to spend more on plant-based meat and dairy alternatives in the future. However, challenges persist, particularly in improving taste and reducing costs, which are the primary barriers to broader adoption.

Taste is the biggest hurdle, with 53% of consumers agreeing that plant-based protein products should taste indistinguishable from meat. Cost is the second most significant factor, deterring both new and repeat purchases. Notwithstanding these obstacles, the market for plant-based meat has tremendous growth potential due to consumer demand, technological developments, and increased investment.

Plant-based meat brings goods closer to the sensory parity of traditional meats and lowers their cost.

Additionally, some consumers are concerned about the perceived "processedness" and the number of ingredients in plant-based meats. To address this, educating consumers on the health benefits of plant-based options and maintaining positive health perceptions will be crucial, especially since health is a primary motivator for those who switch to these alternatives [23].

Some businesses invest in businesses that produce alternatives to animal products as a hedge against or prepare for the drop in demand for these items. In recent years, the market has benefited from funding and investments from well-known people, businesses, and financial investors. A substantial increase over the USD 693 million raised in 2019, the Good Food Institute reports that the alternative proteins market received USD 2.9 billion in declared investments in 2022, while plant-based meat, egg, and dairy companies garnered USD 1.9 billion in the same year. Since 2010, businesses that produce plant-based meat, seafood, eggs, and dairy products have attracted USD 6.3 billion in funding [34].

For instance, No Meat Factory, a Canadian company that produces plant-based meat, received USD 42 million in Series B funding in January 2023. Emil Capital Partners and Tengelmann Growth Partners spearheaded the funding round. Meati (US), a producer of plant-based meat, obtained USD 150 million in Series C funding in July 2022 to boost production [34].

Investing in product renovation and innovation is essential to enhancing plant-based meat products, making them more appealing to consumers by improving taste, texture, and appearance. Businesses are concentrating on reformulating products to minimize any background flavors that could change the expected taste of traditional meat or seafood while achieving more intense and authentic flavors. Another crucial area is texture, where attempts are made to create goods with a well-balanced feel that combines the ideal amount of hardness, softness, and a realistic bite-down experience.

Since most products on the market today rely on a small number of proteins, such as wheat and soy, research into new plant protein bases is also in progress. Investigating different proteins may result in better components and better products. Furthermore, novel methods of using cultured and plant-based fats are being investigated to improve plant-based meat substitutes' flavor, texture, softness, and overall mouthfeel. There is also a push to provide better taste and quality goods but with cleaner ingredient labeling and additional health benefits for health-conscious consumers [23].

Universities and research organizations have significant opportunities to advance the alt-meat sector by addressing its key challenges and accelerating product development. To maximize the benefits and minimize potential risks, research could focus on enhancing open-source technology and providing publicly accessible data, which could speed up market entry and scaling. The cultured meat industry, in particular, has common needs, such as developing effective scaffolds and cell culture media and advancing basic crop science to create new plant variants optimized for alt-meat production. Research on the social and economic effects of switching from conventional animal agriculture to alternative meat is also necessary, especially to help rural farmers make fair transitions. Finding and measuring the potential of these communities may include conducting thorough social science research and systems analysis. Agricultural extension services might also play a role in supporting crop transitions.

By offering legal clarity, government organizations might significantly contribute to expanding the alt-meat sector by streamlining the route to market, particularly in areas like labeling and inspection. Governments could also incentivize land transitions through tax credits for rewilding unused land and support research initiatives with public funding, focusing on developing open-source technologies essential for the alt-meat sector. Additionally, reallocating or removing subsidies from traditional animal agriculture could help level the playing field for alt-meat companies. In order to promote fair transitions for farmers, governments should also establish programs that include debt forgiveness, loss compensation, and financing retraining programs. Economic gains may also result from encouraging the construction of facilities for the production of alternative meats in underserved communities. Governments can significantly impact the alt-meat business by providing public financing to promote research programs. $190 million was set aside for this sector's research and development in 2023 alone. Promoting open-access research and development centered on the main obstacles facing plant-based meat should stimulate industry innovation, going beyond the endeavors of individual businesses to group breakthroughs.

Nonprofit organizations (NGOs) can play a pivotal role in transitioning to alt-meat by fostering stakeholder communication and collaboration. NGOs can promote the animal meat and alternative meat industries and assist in navigating complicated problems like labeling by bringing together disparate parties for candid, nonconfrontational discussions. NGOs may also work with government organizations to provide retraining programs and transition cost subsidies to help with rural transitions. Some nonprofits could advocate for an even playing field by representing the collective interests of alt-meat companies in policy, regulation, labeling, and research funding—The Good Food Institute serves as an example. Education and awareness are critical across these roles, with NGOs helping farmers, ranchers, and communities prepare for changes and capitalize on new opportunities [21].

Although NGOs are essential to this change, they must also handle potential conflicts between their mission-driven agendas and broader agricultural goals. Clear communication is essential to ensuring that their efforts do not inadvertently impede outcomes for rural producers. Continuous communication between stakeholders will be essential, as the alt-meat industry develops in order to resolve issues and optimize the potential for everybody involved [35].

The World Economic Forum's Innovation with a Purpose initiative highlights the need for a dynamic innovation ecosystem to scale alternative proteins and feedstocks. Governments, businesses, innovators, and end users should work together in this ecosystem, particularly in underprivileged areas. By encouraging new company models, enhancing financing methods, reducing risks, and bolstering pertinent policies, such a framework can propel market development.

Drawing from the successful CGIAR model, which advanced agricultural research in the 1970s, there is potential to create a modern equivalent focused on protein innovation. A twenty-first-century version, such as a Consultative Group for International Protein Options Acceleration (CGIPA), could establish a global network of technology, policy, and financial hubs tailored to regional needs. By 2030, this strategy would seek to accelerate the shift to a sustainable global protein system by tackling regional and global issues with trade regulations and protein supply chains [29].

Companies should concentrate on these four crucial tactics in order to gain broad adoption of alternative protein products:

1. Understand Mainstream Consumer Needs: Brands must understand mainstream consumers' demands and preferences, including their context and emotional needs when utilizing the product. Offerings can be better tailored to suit customer expectations if it is known where, with whom, and how they use the product. Continual innovation is essential to satisfying changing customer demands. Companies that invest in learning about consumer preferences using cutting-edge research methods, such as sentiment analysis and predictive analytics, are better able to predict consumer wants and market trends.
2. Innovate for Better Taste, Texture, and Price: For the industry to advance, constant innovation is necessary. Even if taste, texture, and cost have significantly improved, many customers still believe that traditional and alternative meats are not comparable. Innovations that enhance flavor and feeling are essential to bridging this gap.

   Innovative food processing techniques like rheometry and extrusion can improve the flavor and texture of plant-based meat substitutes. By using these technologies, producers can produce goods that more nearly resemble the fibrous structure of animal proteins, increasing customer acceptance [36]. Companies should aim to match and exceed traditional products in fulfilling consumer needs by continually developing and testing new products.
3. Tailor Messaging to Mainstream Preferences: Focusing on more than just sustainability is necessary for effective communication with mainstream consumers. While avoiding specialized phrases like "vegan" and highlighting the product's sensory appeal, messaging should cater to their individual needs and preferences, such as convenience, quality, and health advantages. Promote the buying of goods that are practical for those with busy schedules. Instead of portraying the product as a niche item, highlight its health benefits, such as its high protein, low sugar, or vitamin content, and position it as a part of a balanced diet.
4. Adopt a Test-and-Learn Approach: To better understand customer behavior and improve their marketing and goods, businesses should employ a test-and-learn approach. Testing various product iterations, packaging, and marketing tactics in simulated or real-world market environments might yield complete insight because traditional consumer surveys might not predict behavior in the real world. Impossible Foods, for example, has improved the nutritional profile and packaging of its ground beef using this type of iterative testing [37].

The demand for different crops used as plant protein sources is predicted to rise as the plant-based meat industry expands, creating new opportunities for farmers. Demand is expected to rise for a number of commodities, such as peas, wheat, and soy, which are necessary ingredients for these products. Those already cultivating or willing to integrate these crops into their rotations could benefit economically. These crops, for example, maybe more profitable and allow farmers to diversify their revenue streams, increasing their financial stability. Furthermore, many leguminous crops that can be included in double-cropping rotations could be used as an extra source of revenue rather than a substitute. The growth of the plant-based sector coincides with the emphasis on agricultural sustainability. Farmers can adopt practices like reducing greenhouse gas emissions from animal husbandry that meet customer demands and promote environmental goals.

Furthermore, the growth of facilities for producing plant-based and cultured meat may lead to the creation of new job possibilities, particularly in rural areas. These establishments, which can be sizable and located in agricultural states, might provide a good substitute for jobs in conventional meat production. The alt-meat industry may offer a substantial increase in worker safety, job standards, and possibilities, as slaughterhouse occupations are frequently linked to hazardous and challenging working conditions, low pay, and exploitation. In addition to creating jobs, investing in these manufacturing facilities might boost local economies by drawing in investment and encouraging innovation in food technology, given the growing demand from consumers for plant-based products [21].

References

1. Good Food Institute India. Insights on packaging and labeling for plant-based meat and dairy in India. https://gfi-india.org/insights-on-packaging-and-labeling-for-plant-based-meat-and-dairy-in-india/ (n.d.).
2. Labelling of Plant-Based Meats: Regulatory Landscape. https://www.ikigailaw.com/article/165/labelling-of-plant-based-meats-regulatory-landscape (2021).
3. Vegconomist. India: Regulations for vegan food. Vegconomist. Retrieved November 29, 2024, from https://vegconomist.com/politics-law/india-regulations-for-vegan-food/ (n.d.).
4. Good Food Institute India. The regulatory status of cultivated meat (p. 7). https://gfi-india.org/wp-content/uploads/2022/03/The-Regulatory-Status-of-Cultivated-Meat.pdf (2022).
5. Good Food Institute India. The regulatory status of cultivated meat (p. 9). https://gfi-india.org/wp-content/uploads/2022/03/The-Regulatory-Status-of-Cultivated-Meat.pdf (2022).
6. Kumar K. Regulations for labelling of plant-based meat products. Corpbiz. https://corpbiz.io/learning/regulations-for-labelling-of-plant-based-meat-products/ (2022).
7. Lähteenmäki-Uutela A, Rahikainen M, Lonkila A, Yang B. Alternative proteins and EU food law. Food Control. 2021;130:108336. https://doi.org/10.1016/j.foodcont.2021.108336.
8. Lexology. France reintroduces bid to ban meat names for plant-based food. Lexology. https://www.lexology.com/library/detail.aspx?g=8c575528-686b-4fbd-ad2a-111c1ead6ae6 (2023).
9. U.S. Food and Drug Administration. FDA modernization efforts: Establishing a unified human foods program and new model for field operations. Retrieved December 1, 2024, from https://www.fda.gov/about-fda/fda-organization/fda-modernization-efforts-establishing-unified-human-foods-program-new-model-field-operations-and (n.d.).

10. Food and Drug Law Institute. Alternative proteins: navigating the maze of U.S. federal and state meat labeling requirements. Retrieved December 1, 2024, from https://www.fdli.org (2023).
11. Congressional Research Service. R47697: Federal regulatory framework for cell-cultured meat and poultry products. Retrieved December 1, 2024, from https://crsreports.congress.gov/product/pdf/R/R47697 (2023).
12. Lee SY, Lee DY, Jeong JW, Kim JH, Yun SH, Mariano EJ, Lee J, Park S, Jo C, Hur SJ. Current technologies, regulation, and future perspective of animal product analogs – a review. Anim Biosci. 2023;36(10):1465–87. https://doi.org/10.5713/ab.23.0029.
13. Ellenberg C. South Korea designates regulation-free zone for cultivated meat to boost production & safety. Cultivated X. https://cultivated-x.com/politics-law/south-korea-regulation-free-zone-cultivated-meat-production-safety/ (2024).
14. Whittaker L. Overcoming obstacles in cultivated meat: cost, scalability, and consumer acceptance. https://insights.figlobal.com/innovation-technology/overcoming-obstacles-in-cultivated-meat-cost-scalability-and-consumer-acceptance (2024).
15. China. Food and Agriculture Organization of the United Nations. https://www.fao.org/nutrition/education/food-dietary-guidelines/regions/countries/china/en/ (n.d.).
16. Novel Food Approval Process in Singapore. https://www.freyrsolutions.com/blog/regulatory-practices novel food approval-process-in-singapore (2024).
17. Whitworth J. Revised food safety law progresses in Singapore | Food Safety News. Food Safety News. https://www.foodsafetynews.com/2024/11/revised-food-safety-law-progresses-in-singapore/ (2024).
18. Jones D. The complex labeling landscape for plant-based meat alternatives (2020).
19. Vegconomist. Nravigating the challenges and opportunities of plant-based meat alternatives. Vegconomist. Retrieved December 1, 2024, from https://vegconomist.com/guest-posts/guest-post-navigating-challenges-opportunities-plant-based-meat-alternatives/ (n.d.).
20. Andreani G, Sogari G, Marti A, Froldi F, Dagevos H, Martini D. Plant-based meat alternatives: technological, nutritional, environmental, market, and social challenges and opportunities. Nutrients. 2023;15(2):452. https://doi.org/10.3390/nu15020452.
21. Newton P, Blaustein-Rejto D. Social and economic opportunities and challenges of plant-based and cultured meat for rural producers in the US. Front Sustain Food Syst. 2021;5 https://doi.org/10.3389/fsufs.2021.624270.
22. Weinrich R. Opportunities for the adoption of health-based sustainable dietary patterns: a review on consumer research of meat substitutes. Sustainability. 2019;11(15):4028. https://doi.org/10.3390/su11154028.
23. Ittoop A. Outlook: the state of the plant-based meat category. The Good Food Institute. https://gfi.org/blog/2023-outlook-the-state-of-the-plant-based-meat-category/ (2023).
24. PPTI News. The price gap between Plant-Based and conventional meat is shrinking, new analysis finds | PPTI News. https://www.proteinproductiontechnology.com/post/the-price-gap-between-plant-based-and-conventional-meat-is-shrinking-new-analysis-finds (n.d.).
25. The Vegan Business Magzine. https://vegconomist.com/studies-and-numbers/report-reveals-plant-based-meat-sales-surge-price-parity-items-australian-market/. In https://vegconomist.com/studies-and-numbers/report-reveals-plant-based-meat-sales-surge-price-parity-items-australian-market/ (n.d.).
26. GFI. Environmental impacts of alternative proteins | GFI. The Good Food Institute. https://gfi.org/resource/environmental-impacts-of-alternative-proteins/ (n.d.).
27. Williams A. Lab meat? Synthetic meat? Cultivated meat? 11 things you need to know. GFI Europe. https://gfieurope.org/lab-meat-synthetic-meat-cultivated-meat-11-things-you-need-to-know/ (2024).
28. Ngah MK, Chriki S, Ellies-Oury M, Liu J, Hocquette J. Consumer perception of "artificial meat" in the educated young and urban population of Africa. Front Nutr. 2023;10 https://doi.org/10.3389/fnut.2023.1127655.

29. World Economic Forum, Waughray D, De Cleene S, Sweet L. Meat: the future – a roadmap for delivering 21st-century protein. World Economic Forum; 2019. https://www3.weforum.org/docs/WEF_White_Paper_Roadmap_Protein.pdf.
30. The Good Food Institute. The state of global policy on alternative proteins. The Good Food Institute. https://gfi.org/resource/alternative-proteins-state-of-global-policy/ (2024).
31. Garza F, Garza F. Plant-based meat needs government support to scale up, but a culture war stands in the way. Grist. https://grist.org/food-and-agriculture/plant-based-meat-needs-government-support-to-scale-up-but-a-culture-war-stands-in-the-way/ (2024).
32. Cardenas J. Governments have put $1 billion toward making better meat alternatives. A lot more is needed. Vox. https://www.vox.com/future-perfect/23849473/cell-cultivated-meat-impossible-beyond-alternatives-vegan-investment-report-infrastructure (2023).
33. Ispp & Ispp. Uncovering change potential in India's public policy. Indian School of Public Policy. https://www.ispp.org.in/exploring-the-potential-for-change-through-indias-public-policy/ (2023).
34. Meticulous Research. Plant-based food market by type (Dairy alternatives, plant-based meat, meals, confectionery, beverages, egg alternatives, seafood), source (Soy, wheat, pea, rice), Distribution channel (B2B, B2C [Convenience store, online retail]) – Global forecast to 2031. https://www.meticulousresearch.com/product/plant-based-food-market-5108 (n.d.).
35. Laestadius LI, Neff RA, Barry CL, Frattaroli S. "We don't tell people what to do": an examination of the factors influencing NGO decisions to campaign for reduced meat consumption in light of climate change. Glob Environ Chang. 2014;29:32–40. https://doi.org/10.1016/j.gloenvcha.2014.08.001.
36. Manager KCMD. Overcoming consumer hesitation: making plant-based meats more accessible. Examining Food. https://www.thermofisher.com/blog/food/overcoming-consumer-hesitation-making-plant-based-meats-more-accessible/ (2024).
37. Taking Alternative Proteins Mainstream. Boston Consulting Group. https://www.bcg.com/publications/2023/taking-alternative-protein-trends-mainst ream (2023).

Chapter 9
Future Directions and Outlook

The field of meat substitute technologies is advancing quickly, fueled by technological innovations, increasing environmental concerns, and changing consumer preferences. Future developments in this sector are likely to transform food production even further, improving sustainability, nutrition, and sensory experiences. Speculating about these advancements entails exploring emerging technologies, overcoming scaling challenges, and considering their broader societal impacts.

One of the most promising areas for future development is cellular agriculture. Cultured meat, which is produced by cultivating animal cells in bioreactors, has already shown the potential to address the ethical and environmental issues associated with conventional livestock farming. Looking ahead, optimizing cell lines and growth media will be crucial for enhancing scalability and reducing production costs. The development of serum-free media, which contain plant-based or recombinant growth factors, could significantly improve the sustainability of cultured meat production. Furthermore, advancements in bioreactor design and automation may allow for higher yields and more consistent product quality. The integration of artificial intelligence (AI) and machine learning (ML) in cellular agriculture could also help refine processes by predicting optimal conditions for cell growth and resource utilization.

The development of edible scaffolds represents a significant advancement in cultured meat technology. Future scaffolds may use bioengineered materials that mimic the texture and marbling of different types of meat. For instance, 3D bioprinting could allow for the precise layering of muscle and fat cells, creating products that are indistinguishable from traditional meat cuts. This method could also enable the customization of nutritional profiles to meet specific dietary needs or preferences, such as increasing omega-3 fatty acids or reducing saturated fat content. Plant-based meat substitutes are also expected to see substantial innovation. Current challenges, such as replicating the complexity of whole muscle cuts and achieving cost parity with animal-derived products, are likely to be overcome through advancements in

S. Ray, S. Suri, *The Sustainable Plate: Unveiling the Science of Meat Substitutes and Impact on Global Health*, https://doi.org/10.1007/978-3-031-94604-2_9

processing technologies. High-moisture extrusion, a widely used technique, may undergo further refinement to produce more realistic fibrous textures and juiciness. Additionally, hybrid products that combine plant-based ingredients with cultured cells or fermentation-derived components could emerge, taking advantage of multiple technologies to deliver better taste, texture, and nutritional value.

Fermentation technologies are poised to play a crucial role in future advancements. Precision fermentation, which utilizes microorganisms to produce specific proteins, fats, or flavor compounds, is expected to greatly expand its capabilities. For example, custom-engineered microbes could create novel bioactive compounds or improve the nutritional content of meat substitutes. These innovations may also make plant-based products more appealing by replicating the complex flavors and aromas associated with animal-derived meat. The integration of computational tools, such as artificial intelligence (AI) and machine learning (ML), will further accelerate innovation across all areas of meat substitute technologies. Predictive modeling could optimize ingredient combinations, processing conditions, and product formulations, significantly reducing both the time and cost associated with product development. Additionally, blockchain and other digital technologies could enhance supply chain transparency, addressing consumer concerns about sustainability and authenticity. As these advancements emerge, regulatory frameworks and public perception will be critical to their success. Robust safety evaluations and clear labeling will be essential for building consumer trust. Furthermore, ensuring equitable access to these technologies will be vital for maximizing their global impact, particularly in tackling food security challenges in underserved regions.

9.1 The Possible Impact of Meat Substitutes on the Global Food System and Human Health

The rise of meat substitutes marks an important development in the global food landscape. These alternatives, which include plant-based proteins, cultured meat, and fermentation-derived products, have the potential to transform food production and consumption. This shift is largely driven by increasing concerns about the environmental, ethical, and health impacts of traditional meat production. As more people adopt meat substitutes, it becomes increasingly important to understand their potential effects on the global food system and on human health.

The global food system faces significant challenges, including the need to feed a growing population, reduce greenhouse gas emissions, and minimize resource depletion. Conventional meat production is resource-intensive, requiring large amounts of land, water, and feed. It also contributes substantially to deforestation, biodiversity loss, and climate change. According to the United Nations Food and Agriculture Organization (FAO), livestock production accounts for approximately 14.5% of global greenhouse gas emissions. Meat substitutes offer a promising solution to these environmental issues. Plant-based meat alternatives, made from soy, pea, and wheat proteins, generally have a smaller environmental footprint compared to conventional meat. Life cycle analyses indicate that producing plant-based alternatives can reduce greenhouse gas emissions by up to 90%, land use by up to 93%, and water use by up to 99% when compared to beef production. Additionally, cultured meat—meat grown from animal cells in a lab—has the potential to significantly lower resource use and emissions, although its energy requirements are still being optimized.

The adoption of meat substitutes could also help improve global food security. By moving away from resource-intensive meat production, we could free up agricultural land for growing crops that are directly consumed by humans, potentially allowing us to feed more people with the same resources. However, this transition needs to be managed carefully to avoid unintended consequences, such as displacing small-scale livestock farmers or worsening economic inequalities in rural communities.

The nutritional profile of meat substitutes plays a significant role in their potential impact on human health. Many plant-based products aim to replicate the taste, texture, and nutritional content of animal meat, often providing similar levels of protein. However, the nutritional quality of these substitutes can vary based on their formulation. Some meat substitutes are fortified with essential nutrients such as iron, vitamin B12, and omega-3 fatty acids to counteract potential deficiencies in plant-based diets. Despite these advantages, there are ongoing concerns about the health implications of highly processed meat substitutes. Many of these products contain high levels of sodium, saturated fats, and additives to enhance flavor and texture, which might pose health risks if consumed in excess. Research indicates that a diet rich in minimally processed plant-based foods—such as fruits, vegetables, legumes, and whole grains—is linked to a lower risk of chronic diseases, including cardiovascular disease, type 2 diabetes, and certain types of cancer. Promoting the development of minimally processed alternatives could better align meat substitutes with public health objectives.

Cultured meat, while less processed than some plant-based alternatives, faces its own nutritional challenges. The nutrient profile of cultured meat depends on the growth medium, and additives used during production. Although cultured meat eliminates concerns regarding antibiotics and zoonotic diseases linked to livestock farming, its long-term health effects need further investigation through clinical studies and real-world consumption data.

The ethical appeal of meat substitutes lies in their ability to reduce animal suffering and align with changing consumer values. The rise of flexitarian, vegetarian, and vegan diets reflects an increasing awareness of animal welfare issues and the environmental impact of meat consumption. Meat substitutes provide a way for individuals to align their dietary choices with their ethical beliefs without sacrificing culinary satisfaction. However, cultural acceptance remains a significant challenge. In many communities, meat is deeply rooted in traditions and social practices. Successfully integrating meat substitutes into various culinary traditions will require culturally sensitive product development and marketing strategies. In regions where meat consumption symbolizes wealth and status, it may be necessary to address perceptions and promote the benefits of meat alternatives through targeted education and outreach efforts.

The global market for meat substitutes is expected to experience significant growth, driven by innovation, investment, and increasing consumer demand. According to market research, the plant-based meat sector alone could be valued at over $15 billion by 2027. This growth presents opportunities for job creation, technological advancement, and economic diversification. However, the rise of meat substitutes may disrupt existing supply chains, especially in regions where livelihoods depend on livestock farming. Policymakers must proactively manage this transition by investing in education, training, and infrastructure to help farmers and workers adapt to new roles within a reimagined food system. Public-private partnerships could be essential in promoting sustainable development and ensuring equitable access to the benefits of meat substitutes.

9.2 The Path Forward

To maximize the positive impact of meat substitutes, a collaborative approach involving multiple stakeholders is essential. Governments can play a crucial role by incentivizing research and development, regulating product labeling for transparency, and promoting public awareness about the environmental and health benefits of alternatives. Industry players must focus on innovation to enhance the taste, affordability, and nutritional quality of meat substitutes while also minimizing their environmental impact. Consumers also have a vital role to play by making informed choices and advocating for sustainable practices. The transition to a food system that incorporates meat substitutes is not without challenges, but it presents an unprecedented opportunity to tackle pressing global issues. By reducing the environmental footprint of food production, improving nutritional options, and encouraging ethical consumption, meat substitutes have the potential to transform the global food system and promote better human health. Ensuring that this transformation is inclusive, sustainable, and culturally sensitive will be key to realizing its full potential.

9.3 A Call to Action

Feeding a growing global population while protecting the health of our planet presents significant challenges that require innovative solutions, particularly in our food systems. It is essential for policymakers, industry stakeholders, and individuals to explore sustainable alternatives to conventional meat. These substitutes not only address the environmental and health issues associated with meat production but also contribute to a more equitable and sustainable future. This call to action urges all stakeholders to prioritize, promote, and adopt sustainable meat substitutes, recognizing their potential to transform food systems worldwide.

Livestock farming significantly contributes to greenhouse gas emissions, deforestation, and water scarcity. According to the United Nations Food and Agriculture Organization (FAO), livestock accounts for almost 14.5% of global greenhouse gas emissions. The intensive use of land and water resources for producing livestock feed further worsens environmental degradation. Transitioning to plant-based or lab-grown meat alternatives presents a promising solution to mitigate these impacts. Research shows that producing plant-based proteins requires substantially less land and water and emits far fewer greenhouse gases compared to conventional meat. By reducing our reliance on livestock, we can lessen the pressure on ecosystems, conserve biodiversity, and combat climate change.

Excessive consumption of red and processed meat has been linked to chronic health conditions, including cardiovascular diseases, type 2 diabetes, and certain types of cancer. Sustainable meat substitutes made from plant-based ingredients such as soy, peas, and wheat, as well as those developed through cellular agriculture, can provide high-quality protein with fewer health risks. These alternatives are often fortified with essential nutrients, including vitamins, minerals, and omega-3 fatty acids, ensuring that they are nutritionally adequate while eliminating harmful compounds like cholesterol and saturated fats. Embracing these substitutes aligns with global health recommendations to shift toward diets rich in plant-based foods, promoting improved population health and reducing healthcare burdens.

The global market for sustainable meat substitutes is expected to grow significantly, creating substantial economic opportunities for businesses and governments. Investment in research and development, along with supportive policies, can drive innovation and increase production, making these alternatives more accessible and affordable. Industry stakeholders can benefit from entering this emerging market while also contributing to sustainability goals. For instance, collaborations between food technology companies and traditional meat producers can hasten the transition, resulting in new revenue streams and job creation in the growing sustainable protein sector.

Policymakers play a critical role in facilitating the adoption of sustainable meat substitutes. By implementing supportive regulations, governments can create an enabling environment for innovation and market growth. Key actions include the following:

- Incentivizing Research and Development: Allocate funding for the development of novel meat substitutes, including plant-based and lab-grown options.
- Implementing Subsidies and Tax Breaks: Provide financial incentives for companies producing sustainable protein sources and disincentivize unsustainable livestock farming practices.
- Setting Sustainable Procurement Standards: Integrate sustainable protein options into public procurement policies for schools, hospitals, and government institutions.
- Raising Public Awareness: Launch campaigns to educate citizens on the environmental and health benefits of meat substitutes, fostering acceptance and demand.

The private sector has the potential to transform the food landscape through innovation and a focus on consumer needs. Companies should prioritize developing affordable, tasty, and nutritionally adequate meat substitutes that meet diverse dietary requirements and cultural preferences. Transparent labeling, strong supply chains, and sustainable packaging are essential for building consumer trust. Additionally, partnerships with retailers and food service providers can improve accessibility, making sustainable options the default choice for consumers.

Individuals play a crucial role in facilitating the transition to more sustainable eating habits through conscious dietary choices and advocacy for systemic change. Embracing flexitarian, vegetarian, or plant-based diets can significantly lower one's environmental impact and enhance personal health. Consumers can support companies that prioritize sustainability and push for a greater availability of meat substitutes in local markets. Social media and community platforms can further amplify these efforts, inspiring others to adopt sustainable eating practices.

The transition to sustainable meat substitutes requires a collaborative effort from governments, industries, and individuals. It is essential to create platforms that bring together researchers, policymakers, businesses, and civil society organizations to address obstacles and implement solutions effectively. By sharing knowledge, resources, and best practices, stakeholders can accelerate progress toward a sustainable and resilient food system.

Adopting sustainable meat substitutes is a critical step in achieving global sustainability goals, including climate mitigation, improved health outcomes, and economic resilience. This is a call to action that highlights the shared responsibility of policymakers, industry stakeholders, and individuals in supporting this transition. By embracing innovation, fostering collaboration, and prioritizing sustainability, we can build a healthier, more equitable, and environmentally sustainable future for generations to come. The time to act is now—let us lead the way toward a transformative food system that nourishes both people and the planet.

Chapter 10
Conclusion

As the global food system faces increasing challenges from environmental degradation, population growth, and the rise of noncommunicable diseases (NCDs), meat substitutes have emerged as a vital, sustainable, and health-conscious alternative. The previous chapters provided a thorough exploration of the science, market dynamics, and transformative potential of both plant-based and cell-based meat alternatives. This conclusion synthesizes the key findings, highlights the scientific advancements that support meat substitutes, and discusses their significant implications for global health and sustainability.

Traditional animal-based meat production is a significant contributor to greenhouse gas emissions (GHGe), deforestation, water scarcity, and biodiversity loss. Studies highlighted that livestock farming accounts for nearly 57% of food-related GHGe, making a compelling case for reducing reliance on animal-based protein. Plant-based meat alternatives (PBMAs) and cultured meat (CM) require significantly less land, water, and energy, offering a viable path toward mitigating climate change. Red and processed meat consumption has been linked to heightened risks of cardiovascular diseases, colorectal cancer, type 2 diabetes, and other chronic health conditions. In contrast, meat substitutes provide comparable protein content with lower saturated fats, higher fiber, and fewer health risks. For instance, substituting red meat with plant-based proteins or legumes has been shown to reduce the risk of type 2 diabetes by 30% and improve gut health, highlighting the potential for PBMAs and CM to contribute to healthier dietary patterns.

The success of meat substitutes hinges on consumer perception, affordability, and availability. While younger consumers and flexitarians are more receptive to these alternatives, challenges such as taste, texture, cultural preferences, and price disparities remain. Overcoming these barriers will require continued innovation, effective marketing, and policy interventions to make meat substitutes more accessible and appealing to a broader audience. The global meat substitute market is projected to grow from $7.9 billion in 2022 to $15.7 billion by 2027, driven by

S. Ray, S. Suri, *The Sustainable Plate: Unveiling the Science of Meat Substitutes and Impact on Global Health*, https://doi.org/10.1007/978-3-031-94604-2_10

increased consumer awareness and technological advancements. Countries like Singapore, Japan, and the United States have implemented regulatory frameworks to support the development and commercialization of meat substitutes, underscoring the importance of public-private partnerships in fostering innovation.

Advances in food science, such as high-moisture extrusion, precision fermentation, and 3D printing, have enabled the production of PBMAs and CM that closely mimic the sensory attributes of traditional meat. Fungal-based mycoproteins, insect proteins, and recombinant protein additives further diversify the spectrum of sustainable protein sources, enhancing the potential for a global dietary shift.

The emergence of meat substitutes represents a confluence of advancements in cellular agriculture, plant-based food processing, and microbial fermentation. These innovations not only address ecological and ethical concerns but also align with global health priorities. Plant-based meat alternatives, such as those derived from soy, pea protein, and mycoprotein, provide high-quality protein with all essential amino acids. They are also rich in dietary fiber, vitamins, and minerals, supporting cardiovascular health, weight management, and gut health. Studies revealed that replacing just 3% of animal protein with plant protein can reduce overall mortality by 10%, demonstrating the profound health benefits of a plant-based dietary shift. With the Global Burden of Disease (GBD) study identifying red and processed meat as risk factors for NCDs, PBMAs and CM present a promising strategy for addressing these concerns. By reducing saturated fat intake and enhancing dietary diversity, these alternatives can lower the prevalence of ischemic heart disease, type 2 diabetes, and colorectal cancer, contributing to the achievement of Sustainable Development Goal 3 (Good Health and Well-Being).

The scalability of meat substitutes offers a practical solution to the dual challenges of food security and environmental sustainability. As the global population approaches 9 billion by 2050, the demand for protein will double, straining natural resources. PBMAs and CM, with their lower resource intensity, can play a critical role in ensuring equitable access to nutritious food while minimizing ecological impact.

10.1 The Transformative Power of Meat Substitutes in Shaping Sustainable Diets

Meat substitutes are not merely alternatives; they are catalysts for a broader transformation in the global food system. By addressing the interlinked challenges of health, sustainability, and ethics, these innovations pave the way for a more equitable and environmentally conscious world. The Food and Agriculture Organization's (FAO) concept of sustainable diets emphasizes low environmental impact, cultural acceptability, and nutritional adequacy. PBMAs and CM align with this vision, offering diets that respect biodiversity, reduce GHGe, and promote long-term health. Adoption of these diets can significantly contribute to achieving SDG 2 (Zero Hunger) and SDG 13 (Climate Action).

The ethical considerations surrounding animal slaughter and the growing recognition of animal welfare have catalyzed interest in meat substitutes. Cultural acceptance remains a critical factor, with innovations tailored to local cuisines and dietary preferences. By integrating traditional culinary practices with modern technology, meat substitutes can bridge the gap between innovation and cultural heritage. Governments and industry stakeholders must collaborate to create an enabling environment for meat substitutes. Policies such as subsidies for plant-based products, taxation of conventional meat, and educational campaigns can accelerate adoption. Investment in research and development (R&D) is crucial for enhancing the taste, texture, and affordability of these alternatives, ensuring their competitiveness in global markets.

Consumer preferences are dynamic, shaped by factors such as taste, price, and health consciousness. By focusing on these priorities, companies can drive market growth and consumer loyalty. The rise of private-label plant-based products, for instance, demonstrates the potential for affordable alternatives to capture mainstream acceptance. The journey toward a sustainable food system is complex, requiring a holistic approach that integrates scientific innovation, policy support, and consumer engagement. Meat substitutes represent a tangible step toward addressing the pressing challenges of climate change, public health, and food security. Continuous advancements in food technology are essential for refining the production processes of PBMAs and CM. Precision fermentation, scalable bioreactors, and serum-free media are promising avenues for reducing production costs and enhancing product quality.

Transparent labeling, safety assurances, and awareness campaigns can mitigate skepticism and foster consumer confidence in meat substitutes. Highlighting the environmental and health benefits of these products will be critical for driving behavioral change. The international community must prioritize collaborative efforts to establish uniform standards and regulatory frameworks for meat substitutes. Initiatives such as the Japan Association for Cellular Agriculture (JACA) and the EU's novel foods regulation provide models for fostering innovation while ensuring consumer protection. Ensuring that meat substitutes are affordable and accessible to diverse populations is paramount for their widespread adoption. Addressing economic disparities and promoting inclusive policies will be vital for realizing the full potential of these alternatives. The potential of meat substitutes to revolutionize the global food system cannot be overstated. By offering sustainable, ethical, and health-conscious alternatives to conventional meat, they address some of the most pressing challenges of our time. The shift toward PBMAs and CM is not merely a dietary change but a paradigm shift toward a more sustainable and compassionate world.

As consumers, policymakers, and industry leaders embrace the transformative power of meat substitutes, the vision of a food system that harmonizes human health with planetary well-being becomes increasingly attainable. Through continued innovation, collaboration, and commitment to sustainability, meat substitutes hold the promise of reshaping the way we produce and consume food, paving the way for a healthier and more equitable future.

Correction to: The Sustainable Plate: Unveiling the Science of Meat Substitutes and Impact on Global Health

Correction to: Chapters 1 and 2 in: S. Ray, S. Suri, *The Sustainable Plate: Unveiling the Science of Meat Substitutes and Impact on Global Health*, https://doi.org/10.1007/978-3-031-94604-2

A belated request was received from the authors to remove the below bulletin texts from the chapter opening page of Chaps. 1 and 2. The correction has been carried out in the revised version of the book.

Chapter 1: Pg 1

- Overview of the book's purpose and objectives.
- Exploring the growing interest in meat substitutes and their role in addressing sustainability and health concerns.
- The importance of scientific research in understanding the impact of meat substitutes on global health outcomes.

Chapter 2: Pg 17

- Definition and classification of meat substitutes (1800).
- Exploring various types of meat substitutes, including plant-based, cultured, and novel protein sources (4325).
- The science behind creating realistic textures, flavors, and nutritional profiles in meat substitutes (3055).

The updated versions of these chapters can be found at
https://doi.org/10.1007/978-3-031-94604-2_1
https://doi.org/10.1007/978-3-031-94604-2_2

S. Ray, S. Suri, *The Sustainable Plate: Unveiling the Science of Meat Substitutes and Impact on Global Health*, https://doi.org/10.1007/978-3-031-94604-2_11

Appendix

Recommended Books

1. *Clean Meat: How Growing Meat Without Animals Will Revolutionize Dinner and the World by Paul Shapiro*—A detailed exploration of the cultured meat industry, its challenges, and its promise for the future.
2. *The Fate of Food by Amanda Little*—A fascinating story of the sustainable food revolution, an environmental journalist and professor asks the question: Is the future of food looking bleak—or better than ever?
3. *We Are the Weather: Saving the Planet Begins at Breakfast by Jonathan Safran Foer*—Focuses on dietary choices, particularly reducing meat consumption, as a way to combat climate change.
4. *The Omnivore's Dilemma: A Natural History of Four Meals by Michael Pollan*—Investigates the modern food chain, including ethical and environmental considerations.
5. *Diet for a Hot Planet: The Climate Crisis at the End of Your Fork and What You Can Do About It by Anna Lappé*—Explores how our food choices impact climate change and offers solutions for more sustainable eating.

Documentaries

1. *Cowspiracy: The Sustainability Secret (2014)*—Explores the environmental impacts of animal agriculture and advocates for plant-based diets.
2. *The Game Changers (2018)*—Highlights the health benefits of plant-based diets, featuring elite athletes and medical experts.
3. *Meat the Future (2020)*—Chronicles the rise of the cultured meat industry and its potential to transform the global food system.

S. Ray, S. Suri, *The Sustainable Plate: Unveiling the Science of Meat Substitutes and Impact on Global Health*, https://doi.org/10.1007/978-3-031-94604-2

4. *Food, Inc. (2008)*—A critical look at industrial food production, including meat production, and its impact on health and the environment.
5. *Forks Over Knives (2011)*—Advocates for whole-food, plant-based diets to improve health and prevent chronic diseases.
6. *Eating Our Way to Extinction (2021)*—Examines how food production and consumption contribute to environmental crises and advocates for plant-based solutions.

Websites

1. The EAT-Lancet Commission on Food, Planet, Health (https://eatforum.org/eat-lancet-commission/)
2. The Good Food Institute (https://gfi.org/)
3. World Resources Institute (WRI)—Food Program (https://www.wri.org/food)
4. Meatless Monday (https://www.mondaycampaigns.org/meatless-monday)
5. Plant Based Food Association (https://plantbasedfoods.org/)
6. Future Food Institute (https://futurefoodinstitute.org/)

Organizations

1. Good Food Institute (GFI)—Focuses on promoting plant-based and cultured meat through innovation, research, and advocacy.
2. World Health Organization (WHO)—Provides guidance on nutrition and sustainable development goals, emphasizing the importance of healthy diets.
3. Greenpeace International—Advocates for sustainable food systems and reduced meat consumption to combat climate change.
4. ProVeg International—A food awareness organization working to reduce global meat consumption by 50% by 2040.
5. Food and Agriculture Organization (FAO)—Offers resources and initiatives to promote sustainable food systems and address global food security challenges.
6. The Humane Society International (HSI)—Focuses on animal welfare and advocates for plant-based and cultured meat alternatives.
7. Future Meat Technologies—A leading company in the development of cultured meat products.
8. Impossible Foods and Beyond Meat—Pioneers in the plant-based meat industry, driving consumer awareness and market adoption.

Glossary

1. GDP—Gross domestic product
2. CAGR—Compound annual growth rate
3. GHGe—Greenhouse gas emissions
4. GHG—Greenhouse gas
5. FAO—Food and Agriculture Organization
6. LEAD—Livestock, Environment, and Development Initiative
7. IHD—Ischemic heart disease
8. GBD—Global Burden of Disease
9. DALYs—Disability-adjusted life years
10. T2D—Type 2 diabetes mellitus
11. LCA—Life cycle assessment
12. TVP—Textured vegetable protein
13. CO_2—Carbon dioxide
14. $kgCO_2eq$—Kilograms of carbon dioxide equivalent
15. R&D—Research and development
16. NESR—Nutrition Evidence Systematic Review
17. SCP—Single-cell protein
18. EFSA—European Food Safety Authority
19. WHO—World Health Organization
20. RNA—Ribonucleic acid
21. NFE—Nitrogen-free extract
22. MyBacon—Mycelium-Based Bacon (brand)
23. TM—*Tenebrio molitor* (mealworm)
24. LM—*Locusta migratoria*
25. AD—*Acheta domesticus*
26. *F. venenatum—Fusarium venenatum*
27. Koji—*Aspergillus oryzae* (fungus used for fermentation)
28. 3D—Three-dimensional

S. Ray, S. Suri, *The Sustainable Plate: Unveiling the Science of Meat Substitutes and Impact on Global Health*, https://doi.org/10.1007/978-3-031-94604-2

29. IMP—Inosine monophosphate
30. HME—High-moisture extrusion
31. GMP—Good Manufacturing Practice
32. HACCP—Hazard Analysis and Critical Control Points
33. EPA—Eicosapentaenoic acid
34. DHA—Docosahexaenoic acid
35. PUFA—Polyunsaturated fatty acid
36. MUFA—Monounsaturated fatty acid
37. SFA—Saturated fatty acid
38. MW—Molecular weight
39. pH—Potential of hydrogen (acidity/alkalinity level)
40. Kcal—Kilocalorie
41. ISO—International Organization for Standardization
42. TgNyr−1—Teragrams of nitrogen per year
43. PBMAs—Plant-based meat alternatives
44. PBMA—Plant-based meat alternative
45. $TgCO_2eq$—Teragrams of carbon dioxide equivalent
46. Gt CO_2eq—Gigatonnes of carbon dioxide equivalent
47. FMCG—Fast-moving consumer goods
48. UVM—University of Vermont
49. IIASA—International Institute for Applied Systems Analysis
50. CIAT—International Center for Tropical Agriculture
51. PBAPAs—Plant-based animal product alternatives
52. AHA—American Heart Association
53. LDL—Low-density lipoprotein
54. LDL-C—Low-density lipoprotein cholesterol
55. USD—United States Dollar
56. HPP—High-pressure processing
57. PEF—Pulsed electric field
58. PDCAAS—Protein digestibility-corrected amino acid score
59. IGF-1—Insulin-like growth factor 1
60. SmF—Submerged fermentation
61. SSF—Solid-state fermentation
62. SCM—Surface culture method
63. BCG—Boston Consulting Group
64. PBM—Plant-based meat
65. CBM—Cell-based meat
66. ABM—Animal-based meat
67. MS—Meat substitutes
68. FOP—Front-of-Package (Labels)
69. WTP—Willingness to pay
70. PP—Purchase preference
71. Quorn™—Trademarked brand name for mycoprotein products
72. Promyc™—Mycoprotein ingredient brand by Mycorena
73. ABUNDA—Mycoprotein ingredient brand by ENOUGH

74. FDA—Food and Drug Administration
75. SFA—Singapore Food Agency
76. PBMS—Plant-based meat substitutes
77. GMOs—Genetically modified organisms
78. CM—Cultured meat
79. NIH—National Institutes of Health
80. AARP—American Association of Retired Persons
81. JACA—Japan Association for Cellular Agriculture
82. SDGs—Sustainable Development Goals
83. FSSAI—Food Safety and Standards Authority of India
84. USDA—United States Department of Agriculture
85. EU—European Union
86. ASSOCHAM—Associated Chambers of Commerce and Industry of India
87. EAT-Lancet—EAT-Lancet Commission on Food, Planet, and Health
88. SMART Proteins—Sustainable, Market-Ready, Alternative, Resilient, and Trusted Proteins

FSC
www.fsc.org
MIX
Papier aus verantwortungsvollen Quellen
Paper from responsible sources
FSC® C105338

If you have any concerns about our products,
you can contact us on
ProductSafety@springernature.com

In case Publisher is established outside the EU,
the EU authorized representative is:
Springer Nature Customer Service Center GmbH
Europaplatz 3, 69115 Heidelberg, Germany

Printed by Libri Plureos GmbH
in Hamburg, Germany